Cambridge Tracts in Mathematics
and Mathematical Physics

GENERAL EDITOR
W. V. D. HODGE

No. 24

INVARIANTS OF
QUADRATIC DIFFERENTIAL
FORMS

INVARIANTS OF QUADRATIC DIFFERENTIAL FORMS

BY

OSWALD VEBLEN

CAMBRIDGE
AT THE UNIVERSITY PRESS
1952

PUBLISHED BY THE PRESS SYNDICATE OF THE UNIVERSITY OF CAMBRIDGE
The Pitt Building, Trumpington Street, Cambridge, United Kingdom

CAMBRIDGE UNIVERSITY PRESS
The Edinburgh Building, Cambridge CB2 2RU, UK
40 West 20th Street, New York NY 10011–4211, USA
477 Williamstown Road, Port Melbourne, VIC 3207, Australia
Ruiz de Alarcón 13, 28014 Madrid, Spain
Dock House, The Waterfront, Cape Town 8001, South Africa

http://www.cambridge.org

First Edition 1927
Reprinted 1933, 1952
First paperback edition 2004

A catalogue record for this book is available from the British Library

ISBN 0 521 06673 5 hardback
ISBN 0 521 60484 2 paperback

PREFACE

WHEN I was asked to write this tract I was given the privilege of making any possible use of the tract by my old friend J. E. Wright, which was published under the same title about twenty years ago. In these twenty years, however, so much has happened to change our view of the subject that I am sure Wright would have written an entirely new tract if he had lived—and that is what I have done.

It is not merely that many new discoveries have been made, but since the advent of Relativity the subject has been so much studied and expounded with a view to its applications that it now seems possible to say that certain methods are definitely accepted as of primary interest and certain others left to one side as of less consequence to science as a whole. I have tried to set forth the parts of the subject which are important for the applications as fully as the space available would permit and therefore have been forced to leave out several of the questions which Wright included. I have also tried to make the tract elementary in the sense that fundamental definitions are carefully formulated. This has necessarily made the preliminary part of the book long as compared with the rest, and has crowded out material on the applications of the subject which I wrote with more pleasure than some of the pages actually included. However, there are so many books on Relativity, and doubtless will be so many others applying differential invariant theory to Electromagnetic theory, Dynamics, and Quantum theory that one may perhaps be forgiven for not trying to include the applications in these few pages.

Differential geometry has also been crowded out. It seemed important to illustrate the general ideas by the simple case from which they are generalized, namely, elementary geometry. This left no room for higher differential geometry, not even for a discussion of infinitesimal parallelism. But the geometrical point of view is accessible in several recent books* with which this one is not intended to compete. Its purpose is rather to assist the students of differential geometry and mathematical physics by setting forth the underlying differential invariant theory. So it is not entirely by accident that the book ends with a formula which can be of interest only to a reader who intends to go forward to the problems in which it is used.

My thanks are due to several of my colleagues and students at Princeton who have made helpful suggestions either when reading the manuscript or during my lectures on the subject. I am particularly indebted to Dr J. M. Thomas and Mr M. S. Knebelman who have read the whole of the manuscript, and the proof sheets as well.

* On differential geometry we may mention D. J. Struik, *Mehrdimensionale Differentialgeometrie*, Berlin, 1922; J. A. Schouten, *Der Ricci-Kalkül*, Berlin, 1924; E. Cartan, *La Géométrie des espaces de Riemann*, Paris, 1925; T. Levi-Civita, *Lezioni di Calcolo Differenziale assoluto*, Rome, 1925 (English translation, London, 1927); L. P. Eisenhart, *Riemannian Geometry*, Princeton, 1926: on differential invariants in general, R. Weitzenböck, *Invariantentheorie*, Groningen, 1923.

OSWALD VEBLEN

PRINCETON, N. J.

CONTENTS

CHAPTER I

FORMAL PRELIMINARIES

1. The summation convention.

The theory of differential invariants, like the theory of algebraic invariants, is essentially formal. It has many geometrical and physical applications which have played a large rôle in the development of the theory. Yet, after all, it is the actual formulas which are the essential subject matter of the theory. This, at least, is the point of view which we are adopting at present, and so we shall devote a chapter to questions of notation before we try to say what a differential invariant is.

Recent advances in the theory of differential invariants and the wide use of this theory in physical investigations have brought about a rather general acceptance of a particular type of notation, the essential feature of which is the systematic use of subscripts and superscripts and the resulting abandonment of all sorts of notations for special operations. The only operations for which we shall employ special signs are addition, subtraction, multiplication, division, differentiation and integration. These, with the usual run of symbols for functions and sets of functions, are found to suffice for all our purposes.

We shall follow the usage of Einstein of indicating a summation by means of a repeated index: i.e. any term in which the same index (subscript or superscript) appears twice shall stand for the sum of all terms obtainable by giving the index all possible values. Thus, for example, if i can take on the values* from 1 to n,

$$(1\cdot1) \qquad a_1x^1 + a_2x^2 + \ldots + a_nx^n = a_ix^i.$$

Before the advent of Relativity, this expression would have been written

$$\sum_{i=1}^{n} a_ix^i.$$

The innovation consists simply in leaving off the summation sign whenever an index is repeated. Its only inconvenience arises when

---

* The superscript k in x^k does not mean that x is raised to the power k but is merely an index to distinguish among n variables, $x^1, x^2, \ldots, x^n$.

we wish to speak of the general term in an expression like (1·1) without carrying out the summation. But in the theory of differential invariants this situation arises so rarely that the inconvenience is negligible in comparison with the advantages of the notation.

The repeated index is sometimes called a *dummy* or an *umbral* index because, like a variable of integration, the symbol for it in any expression can be changed without affecting the meaning of the expression. Thus

$$a_i x^i = a_k x^k.$$

We shall have to deal with sets of quantities

(1·2) $$T^{ab...c}_{ij...k}$$

which are in general functions of n variables x^1, x^2, ..., x^n. If there are p superscripts and q subscripts each taking values from 1 to n, the expression (1·2) indicates a set of n^{p+q} quantities. By setting a subscript and a superscript equal to each other and summing according to the summation convention we can get a new set of quantities, for example,

$$P^{b...c}_{j...k} = T^{ab...c}_{aj...k}.$$

This operation is called *contraction* (German, *Verjüngung*).

When we have two sets of quantities and multiply every quantity in one set by every quantity in the other set, for example,

$$P^{ab...c}_{ij...k} Q^{de...f}_{lm...p} = R^{ab...cde...f}_{ij...klm...p},$$

we get a new and more numerous set of quantities of the same type. This operation is called *multiplication*. When we have two sets of quantities indicated by the same numbers of subscripts and of superscripts a new set of quantities of the same type is obtained by *addition*,

$$P^{ab...c}_{ij...k} + Q^{ab...c}_{ij...k} = S^{ab...c}_{ij...k}.$$

As an illustration of these operations we may write the formula for a multiple power series

$$A + A_i x^i + \frac{1}{2} A_{ij} x^i x^j + \frac{1}{3!} A_{ijk} x^i x^j x^k + ...$$

A set of quantities such as (1·2) is said to be *symmetric* in any set of indices provided that the value of the symbol (1·2) is unaltered by any permutation of the indices in question. For example, if

$$\Gamma^i_{jk} = \Gamma^i_{kj},$$

the quantities Γ are *symmetric* in the subscripts. A set of quantities is said to be *alternating* (or antisymmetric or skew-symmetric) in a given set of indices providing it is unchanged by any even permutation* of the set of indices in question and merely changed in sign by any odd permutation of the same indices. For example, if

$$\Gamma^{i}_{jk} = -\Gamma^{i}_{kj},$$

the quantities Γ are alternating in the subscripts.

2. The Kronecker deltas.

The theory of determinants and allied expressions is essentially a theory of alternating sets of quantities, and can be made to depend on certain fundamental alternating sets of quantities which have only the values 0 and $+1$ and -1. These sets of quantities are known as generalized Kronecker deltas because of their analogy with the Kronecker delta which is already well known. The latter is defined as follows:

$$\delta^{i}_{j} = 1, \text{ if } i = j; \text{ and } \delta^{i}_{j} = 0, \text{ if } i \neq j.$$

Hence using the summation convention

(2·1) $\delta^{i}_{i} = n,$

and $\delta^{i}_{j} a_{i} = a_{j}, \text{ and } \delta^{i}_{j} a^{j} = a^{i}.$

If $x^1, x^2, \ldots, x^n$ are independent variables,

(2·2) $\dfrac{\partial x^{i}}{\partial x^{j}} = \delta^{i}_{j}.$

The generalized Kronecker delta has k superscripts and k subscripts, each running from 1 to n, and is alternating both in superscripts and subscripts. It is denoted by

$$\delta^{i_1 i_2 \ldots i_k}_{j_1 j_2 \ldots j_k}.$$

If the superscripts are distinct from each other and the subscripts are the same set of numbers as the superscripts, the value of the symbol is $+1$ or -1 according as an even or an odd permutation is required to arrange the superscripts in the same order as the subscripts; in all other cases its value is 0.

---

* It is proved in books on algebra that any permutation of n objects can be brought about by a finite number of transpositions of pairs of these objects, and that the number of transpositions required to bring about a given permutation is always even or always odd. If this number is even the permutation is said to be even. In the opposite case the permutation is said to be odd.

For example, if $n = 3$ and $k = 2$

$$0 = \delta_{ij}^{11} = \delta_{ij}^{22} = \delta_{11}^{ij} = \delta_{13}^{12}, \text{ etc. } \quad 1 = \delta_{12}^{12} = \delta_{13}^{13} = \delta_{21}^{21}, \text{ etc.}$$

$$-1 = \delta_{21}^{12} = \delta_{31}^{13} = \delta_{12}^{21}, \text{ etc.}$$

Using these symbols, the general formula for any two-rowed determinant formed from the matrix

$$\begin{pmatrix} x^1 x^2 \dots x^n \\ y^1 y^2 \dots y^n \end{pmatrix}$$

is

$$x^i y^j - x^j y^i = \delta_{ab}^{ij} x^a y^b.$$

In like manner we can represent differential expressions which are analogous to determinants. For example, if A_1, A_2, ..., A_n are functions of x^1, x^2, ..., x^n,

$$\frac{\partial A_i}{\partial x^j} - \frac{\partial A_j}{\partial x^i} = \delta_{ij}^{ab} \frac{\partial A_a}{\partial x^b}.$$

The general three-rowed determinant formed from the matrix

$$\begin{pmatrix} A_1 & A_2 & A_3 & \dots & A_n \\ B_1 & B_2 & B_3 & \dots & B_n \\ C_1 & C_2 & C_3 & \dots & C_n \end{pmatrix}$$

is

$$\delta_{ijk}^{abc} A_a B_b C_c = \begin{vmatrix} A_i & A_j & A_k \\ B_i & B_j & B_k \\ C_i & C_j & C_k \end{vmatrix}.$$

If A_i and B_i are analytic functions of x^1, x^2, ..., x^n, we can form determinant-like expressions as follows:

$$\delta_{ijk}^{abc} A_a \frac{\partial B_b}{\partial x^c} = A_i \left(\frac{\partial B_j}{\partial x^k} - \frac{\partial B_k}{\partial x^j} \right) + A_j \left(\frac{\partial B_k}{\partial x^i} - \frac{\partial B_i}{\partial x^k} \right) + A_k \left(\frac{\partial B_i}{\partial x^j} - \frac{\partial B_j}{\partial x^i} \right).$$

If a set of quantities T is symmetric in two or more subscripts

$$(2\cdot3) \qquad \delta_{pq\dots r}^{ij\dots l} T_{ij\dots l}^{ab\dots c} = 0,$$

and if it is symmetric in two or more superscripts an analogous relation holds. If a set of quantities A is alternating in its subscripts, which are k in number,

$$(2\cdot4) \qquad \delta_{pq\dots r}^{ij\dots l} A_{ij\dots l} = k!\, A_{pq\dots r}.$$

3. In studying determinants it is often advantageous to use two other permutation symbols defined as follows:

$$(3\cdot1) \qquad \epsilon^{i_1 i_2 \dots i_n} = \delta_{12\dots n}^{i_1 i_2 \dots i_n} = \delta_{i_1 i_2 \dots i_n}^{12\dots n} = \epsilon_{i_1 i_2 \dots i_n}.$$

Thus ϵ is $+1$ or -1 according as the subscripts or superscripts are obtained from the natural numbers $1\,2\,\ldots\,n$ by an even or an odd permutation; otherwise it is zero. The number of indices on an epsilon is always n. By the definition of a determinant,

$$(3\cdot2) \qquad a = |\, a\,^i_{\ j}\,| = \begin{vmatrix} a^1_1 & a^1_2 & \ldots & a^1_n \\ a^2_1 & a^2_2 & \ldots & a^2_n \\ \vdots & & & \\ a^n_1 & a^n_2 & \ldots & a^n_n \end{vmatrix}$$

$$= \epsilon^{i_1 i_2 \ldots i_n} a^1_{i_1} a^2_{i_2} \ldots a^n_{i_n} = \epsilon_{i_1 i_2 \ldots i_n} a^{i_1}_1 a^{i_2}_2 \ldots a^{i_n}_n.$$

For example, a generalized Kronecker delta is a determinant of the simple Kronecker deltas,

$$(3\cdot3) \qquad \delta^{i_1 i_2 \ldots i_k}_{j_1 j_2 \ldots j_k} = \epsilon_{ab\ldots c}\, \delta^{i_a}_{j_1} \delta^{i_b}_{j_2} \ldots \delta^{i_c}_{j_k}.$$

(Here we are applying the summation convention to indices and subscripts of indices.) From either of the expansions in $(3\cdot2)$ it is evident that interchanging two rows of the determinant changes its sign. Hence for any permutation of rows

$$(3\cdot41) \qquad a\epsilon^{ab\ldots c} = \epsilon^{ij\ldots k} a^a_i a^b_j \ldots a^c_k.$$

Likewise, for any permutation of columns

$$(3\cdot42) \qquad a\epsilon_{ab\ldots c} = \epsilon_{ij\ldots k} a^i_a a^j_b \ldots a^k_c.$$

It is an obvious corollary of these two formulas that $a = 0$ if any two rows or any two columns of the determinant are identical.

The formula for the product of two determinants may be derived as follows:

$$(3\cdot5) \qquad ab = a\epsilon_{ab\ldots c}\, b^a_1 b^b_2 \ldots b^c_n$$

$$= \epsilon_{ij\ldots m} a^i_a a^j_b \ldots a^m_c\, b^a_1 b^b_2 \ldots b^c_n$$

$$= \epsilon_{ij\ldots m} (a^i_a b^a_1)(a^j_b b^b_2) \ldots (a^m_c b^c_n)$$

$$= |\, c^i_j\,|,$$

where

$$(3\cdot6) \qquad c^i_j = a^i_a b^a_j = a^i_1 b^1_j + a^i_2 b^2_j + \ldots + a^i_n b^n_j.$$

The formula for the expansion of a determinant in terms of the elements of the first column and their cofactors may be obtained as follows:

$$(3\cdot7) \qquad a = a^{i_1}_1 \epsilon_{i_1 i_2 \ldots i_n} a^{i_2}_2 \ldots a^{i_n}_n = a^i_1 A^1_i,$$

where

(3·8) $$A^1_i = \epsilon_{ii_2\ldots i_n} a^{i_2}_2 a^{i_3}_3 \ldots a^{i_n}_n.$$

More generally,

(3·9) $$a = a^j_p \epsilon_{i_1 i_2 \ldots i_n} a^{i_1}_1 \ldots a^{i_{p-1}}_{p-1} \delta^{i_p}_j a^{i_{p+1}}_{p+1} \ldots a^{i_n}_n.$$

Hence if we define the cofactor of a^j_p as

(3·10) $$A^p_j = \epsilon_{i_1 i_2 \ldots i_n} a^{i_1}_1 \ldots a^{i_{p-1}}_{p-1} \delta^{i_p}_j a^{i_{p+1}}_{p+1} \ldots a^{i_n}_n,$$

we have

(3·11) $$a^j_q A^p_j = a\delta^p_q.$$

This formula gives the expansion of the determinant in terms of the elements of the pth column if $p = q$. In case $p \neq q$ it gives the theorem that the sum of the products of the elements of one column into the cofactors of another is zero. The corresponding theorems about the expansion in terms of the elements of a row are

(3·12) $$a^j_p A^p_i = a\delta^j_i.$$

Although we have spoken about rows and columns it is clear that the visual representation of a determinant may be left to one side when we are using the present notation. The notation takes the place of these other devices. It is not merely an abbreviation; it is a measure for economy of thought. For by arranging that unessential or routine questions are taken care of automatically it enables us to concentrate attention on the new ideas which we have to meet.

4. Linear equations.

To solve a set of linear equations

(4·1)
$$a^1_1 x^1 + a^1_2 x^2 + \ldots + a^1_n x^n = b^1,$$
$$a^2_1 x^1 + a^2_2 x^2 + \ldots + a^2_n x^n = b^2,$$
$$\vdots \qquad\qquad \vdots \qquad\quad \vdots$$
$$a^n_1 x^1 + a^n_2 x^2 + \ldots + a^n_n x^n = b^n,$$

or, as we prefer to write them,

(4·2) $$a^i_j x^j = b^i,$$

we multiply (4·2) by A^k_i and sum with respect to i,

$$a^i_j A^k_i x^j = b^i A^k_i.$$

Using (3·11) this reduces to

$$a\delta^k_j x^j = b^i A^k_i,$$

or, in case $a \neq 0$,

$$(4 \cdot 3) \qquad x^k = \frac{b^i A_i^k}{a},$$

which is Cramer's rule for the solution of linear equations.

5. Functional determinants.

In our work the determinant which appears most frequently is the Jacobian of n functions of n variables

$$y^i (x^1, x^2, \ldots, x^n),$$

which is defined by the equation

$$(5 \cdot 1) \qquad \frac{\partial (y^1, y^2, \ldots, y^n)}{\partial (x^1, x^2, \ldots, x^n)} = \left| \frac{\partial y}{\partial x} \right|$$

$$= \epsilon^{i_1 i_2 \ldots i_n} \frac{\partial y^1}{\partial x^{i_1}} \frac{\partial y^2}{\partial x^{i_2}} \cdots \frac{\partial y^n}{\partial x^{i_n}}.$$

For n functions $z^i (y^1, y^2, \ldots, y^n)$ a fundamental theorem on partial differentiation states that

$$(5 \cdot 2) \qquad \frac{\partial z^i}{\partial x^j} = \frac{\partial z^i}{\partial y^k} \frac{\partial y^k}{\partial x^j}.$$

Hence by the theorem on multiplication of determinants (3·6) the functional determinants satisfy the equation

$$(5 \cdot 3) \qquad \left| \frac{\partial z}{\partial x} \right| = \left| \frac{\partial z}{\partial y} \right| \left| \frac{\partial y}{\partial x} \right|.$$

In case the functions z^i are such that

$$z^i (y^1, y^2, \ldots, y^n) = x^i,$$

(5·2) becomes

$$(5 \cdot 4) \qquad \frac{\partial x^i}{\partial y^k} \frac{\partial y^k}{\partial x^j} = \delta_j^i,$$

and (5·3) reduces to

$$(5 \cdot 5) \qquad \left| \frac{\partial y}{\partial x} \right| = \frac{1}{\left| \dfrac{\partial x}{\partial y} \right|}.$$

For a fixed value of j, (5·4) may be regarded as a set of n linear equations for the determination of n unknowns, $\partial y^k / \partial x^j$, the coefficients being the n^2 quantities $\partial x^i / \partial y^k$. Solving these equations according to § 4, we find

$$(5 \cdot 6) \qquad \frac{\partial y^i}{\partial x^j} = \frac{\text{cofactor of } \dfrac{\partial x^j}{\partial y^i} \text{ in } \left| \dfrac{\partial x}{\partial y} \right|}{\left| \dfrac{\partial x}{\partial y} \right|}$$

Other formulas about functional determinants are:

(5·7)
$$\delta_{ij\ldots k}^{ab\ldots c}\left|\frac{\partial y}{\partial x}\right| = \frac{\partial\,(y^a y^b \ldots y^c)}{\partial\,(x^i x^j \ldots x^k)},$$

and

(5·8)
$$\delta_{ij\ldots k}^{ab\ldots c} = \frac{\partial\,(x^a x^b \ldots x^c)}{\partial\,(x^i x^j \ldots x^k)}.$$

6. By partial differentiation the formulas (5·4) give rise to the following formulas which are often useful:

(6·1)
$$\frac{\partial^2 y^i}{\partial x^a \partial x^b}\frac{\partial x^a}{\partial y^j} + \frac{\partial^2 x^a}{\partial y^j \partial y^k}\frac{\partial y^i}{\partial x^a}\frac{\partial y^k}{\partial x^b} = 0;$$

(6·2)
$$\frac{\partial^2 y^i}{\partial x^a \partial x^b}\frac{\partial x^a}{\partial y^j}\frac{\partial x^b}{\partial y^k} + \frac{\partial^2 x^a}{\partial y^j \partial y^k}\frac{\partial y^i}{\partial x^a} = 0;$$

(6·3)
$$\frac{\partial^3 y^i}{\partial x^a \partial x^b \partial x^c}\frac{\partial x^a}{\partial y^j} + \frac{\partial^2 y^i}{\partial x^a \partial x^b}\frac{\partial^2 x^a}{\partial y^j \partial y^k}\frac{\partial y^k}{\partial x^c} + \frac{\partial^3 x^a}{\partial y^j \partial y^k \partial y^l}\frac{\partial y^i}{\partial x^a}\frac{\partial y^k}{\partial x^b}\frac{\partial y^l}{\partial x^c}$$
$$+ \frac{\partial^2 x^a}{\partial y^j \partial y^k}\frac{\partial^2 y^i}{\partial x^a \partial x^c}\frac{\partial y^k}{\partial x^b} + \frac{\partial^2 x^a}{\partial y^j \partial y^k}\frac{\partial y^i}{\partial x^a}\frac{\partial^2 y^k}{\partial x^b \partial x^c} = 0;$$

and so on.

7. Derivative of a determinant.

In case the elements of a determinant are functions of $(x^1, x^2, \ldots, x^n)$ we have by differentiating (3·2)

(7·1)
$$\frac{\partial a}{\partial x^j} = \epsilon_{i_1 i_2 \ldots i_n}\left(\frac{\partial a_1^{i_1}}{\partial x^j} a_2^{i_2} \ldots a_n^{i_n} + a_1^{i_1}\frac{\partial a_2^{i_2}}{\partial x^j} \ldots a_n^{i_n} + \ldots\right)$$
$$= \frac{\partial a_1^{i_1}}{\partial x^j} A_{i_1}^1 + \frac{\partial a_2^{i_2}}{\partial x^j} A_{i_2}^2 + \ldots$$
$$= \frac{\partial a_c^b}{\partial x^j} A_b^c.$$

In case the determinant in question is the Jacobian of a transformation of coordinates, (5·1), (7·1) reduces to

(7·2)
$$\frac{\partial}{\partial x^j}\left|\frac{\partial y}{\partial x}\right| \Big/ \left|\frac{\partial y}{\partial x}\right| = \frac{\partial^2 y^b}{\partial x^j \partial x^c}\frac{\partial x^c}{\partial y^b}.$$

8. Numerical relations.

The permutation symbols satisfy a number of numerical relations which are easily verified by counting the number of terms which appear in the various sums. Thus we have

$$\delta^{ij}_{pj} = (n-1)\delta^i_p, \quad \text{and} \quad \delta^{ij}_{ij} = n\,(n-1),$$

(8·1) $$\delta^{i_1 i_2 \ldots i_r i_{r+1} \ldots i_k}_{j_1 j_2 \ldots j_r i_{r+1} \ldots i_k} = \frac{(n-r)!}{(n-k)!}\,\delta^{i_1 i_2 \ldots i_r}_{j_1 j_2 \ldots j_r},$$

(8·2) $$\delta^{i_1 i_2 \ldots i_k}_{i_1 i_2 \ldots i_k} = \frac{n!}{(n-k)!},$$

(8·3) $$\epsilon^{ab \ldots m}\,\epsilon_{ab \ldots m} = n!,$$

(8·4) $$\epsilon^{i_1 \ldots i_k i_{k+1} \ldots i_n}\,\epsilon_{j_1 \ldots j_k i_{k+1} \ldots i_n} = (n-k)!\,\delta^{i_1 \ldots i_k}_{j_1 \ldots j_k},$$

(8·5) $$\epsilon^{i_1 \ldots i_k i_{k+1} \ldots i_n}\,\delta^{j_{k+1} \ldots j_n}_{i_{k+1} \ldots i_n} = (n-k)!\,\epsilon^{i_1 \ldots i_k j_{k+1} \ldots j_n},$$

(8·6) $$\delta^{i_1 \ldots i_k i_{k+1} \ldots i_r}_{j_1 \ldots j_k j_{k+1} \ldots j_r}\,\delta^{j_{k+1} \ldots j_r}_{p_{k+1} \ldots p_r} = (r-k)!\,\delta^{i_1 \ldots i_k i_{k+1} \ldots i_r}_{j_1 \ldots j_k p_{k+1} \ldots p_r},$$

(8·7) $$\delta^{i_1 \ldots i_k i_{k+1} \ldots i_r}_{j_1 \ldots j_k j_{k+1} \ldots j_r}\,\delta^{j_{k+1} \ldots j_r}_{i_{k+1} \ldots i_r} = \frac{(n-k)!}{(n-r)!}\,(r-k)!\,\delta^{i_1 \ldots i_k}_{j_1 \ldots j_k},$$

(8·8) $$\epsilon_{i \ldots jk \ldots lm} = (-1)^p\,\epsilon_{i \ldots jmk \ldots l},$$

if p is the number of indices $k \ldots l$.

9. Minors, cofactors, and the Laplace expansion.

The k-rowed *minors* of a determinant a are defined by the formula

(9·1) $$a^{i_1 i_2 \ldots i_k}_{j_1 j_2 \ldots j_k} = \delta^{i_1 i_2 \ldots i_k}_{p_1 p_2 \ldots p_k}\,a^{p_1}_{j_1}\,a^{p_2}_{j_2} \ldots a^{p_k}_{j_k}$$
$$= \delta^{p_1 p_2 \ldots p_k}_{j_1 j_2 \ldots j_k}\,a^{i_1}_{p_1}\,a^{i_2}_{p_2} \ldots a^{i_k}_{p_k}.$$

Thus the one-rowed minors are the elements a^i_j themselves, the two-rowed minors are the determinants,

$$\begin{vmatrix} a^{i_1}_{j_1} & a^{i_2}_{j_1} \\ a^{i_1}_{j_2} & a^{i_2}_{j_2} \end{vmatrix},$$

and so on. The determinant is given by

(9·2) $$a = \frac{1}{n!}\,a^{i_1 i_2 \ldots i_n}_{i_1 i_2 \ldots i_n},$$

and we also have

(9·3) $$a\delta^{i_1 \ldots i_k}_{j_1 \ldots j_k} = \frac{1}{(n-k)!}\,a^{i_1 \ldots i_k i_{k+1} \ldots i_n}_{j_1 \ldots j_k i_{k+1} \ldots i_n}.$$

The *cofactor* of the k-rowed minor (9·1) is the determinant

(9·4) $$A^{j_1 j_2 \ldots j_k}_{i_1 i_2 \ldots i_k} = \frac{1}{(n-k)!}\,\delta^{j_1 j_2 \ldots j_n}_{i_1 i_2 \ldots i_n}\,a^{i_{k+1}}_{j_{k+1}} \ldots a^{i_n}_{j_n}$$
$$= \frac{1}{((n-k)!)^2}\,\delta^{j_1 \ldots j_n}_{i_1 \ldots i_n}\,a^{i_{k+1} \ldots i_n}_{j_{k+1} \ldots j_n},$$

which is equivalent to (3·10) if $k = 1$.

Applying some of the formulas of § 8, we find

$$(9\cdot5)\quad a^{i_1\ldots i_k}_{j_1\ldots j_k}A^{j_1\ldots j_k}_{q_1\ldots q_k}=\delta^{p_1\ldots p_k}_{j_1\ldots j_k}a^{i_1}_{p_1}\ldots a^{i_k}_{p_k}\frac{1}{(n-k)!}\delta^{j_1\ldots j_k\ldots j_n}_{q_1\ldots q_k\ldots q_n}a^{q_{k+1}}_{j_{k+1}}\ldots a^{q_n}_{j_n}$$

$$=\frac{k!}{(n-k)!}\delta^{p_1\ldots p_k j_{k+1}\ldots j_n}_{q_1\ldots q_k q_{k+1}\ldots q_n}a^{i_1}_{p_1}\ldots a^{i_k}_{p_k}a^{q_{k+1}}_{j_{k+1}}\ldots a^{q_n}_{j_n}$$

$$=\frac{k!}{(n-k)!}a^{i_1\ldots i_k q_{k+1}\ldots q_n}_{q_1\ldots q_k q_{k+1}\ldots q_n}$$

$$=k!\,a\delta^{i_1\ldots i_k}_{q_1\ldots q_k},$$

a formula which includes the Laplace expansion, and therefore, also (3·11), as a special case. Similarly we can get

$$(9\cdot6)\qquad \delta^{j_1\ldots j_n}_{s_1\ldots s_n}a^{i_1\ldots i_k}_{j_1\ldots j_k}a^{i_{k+1}\ldots i_n}_{j_{k+1}\ldots j_n}=k!\,(n-k)!\,a^{i_1\ldots i_n}_{s_1\ldots s_n},$$

which also includes the Laplace expansion.

10. For reference later on it will be convenient to write out the formulas for the minors and cofactors of a matrix

$$(10\cdot1)\qquad \|\,g_{ij}\,\| = \begin{pmatrix} g_{11} & g_{12} & \cdots & g_{1n} \\ g_{21} & g_{22} & \cdots & g_{2n} \\ \vdots & \vdots & & \vdots \\ g_{n1} & g_{n2} & \cdots & g_{nn} \end{pmatrix},$$

in a notation which is slightly different in appearance from that just given, though in fact equivalent to it. The k-rowed minors are

$$(10\cdot2)\qquad g_{ab\ldots c;\,ij\ldots m}=\delta^{pq\ldots r}_{ab\ldots c}\,g_{pi}g_{qj}\ldots g_{rm}.$$

According to this definition the one-rowed minors are the elements g_{ij} themselves. The two-rowed minors satisfy the conditions

$$(10\cdot3)\qquad\qquad g_{ab;\,ij}=-g_{ba;\,ij}$$
$$=-g_{ab;\,ji},$$

and the k-rowed minors satisfy analogous relations. In the important special case in which

$$(10\cdot4)\qquad\qquad g_{ij}=g_{ji},$$

we also have

$$(10\cdot5)\qquad\qquad g_{ab;\,ij}=g_{ij;\,ab}.$$

The cofactors of order k are given by the formulas

$$(10\cdot6)\quad G^{a_1\ldots a_k;\,i_1\ldots i_k}=\frac{1}{(n-k)!}\,\epsilon^{a_1\ldots a_n}\epsilon^{i_1\ldots i_n}g_{a_{k+1}\,i_{k+1}}\ldots g_{a_n\,i_n}.$$

So in particular, the cofactor of g_{ai} is

$$(10 \cdot 7) \qquad G^{a;i} = \frac{1}{(n-1)!} \epsilon^{aa_2 \ldots a_n} \epsilon^{ii_2 \ldots i_n} g_{a_2 i_2} \ldots g_{a_n i_n}.$$

The Laplace expansion becomes

$$(10 \cdot 8) \qquad G^{ab \ldots c; ij \ldots k} g_{ab \ldots c; st \ldots u} = r! \, g \delta^{ij \ldots k}_{st \ldots u},$$

in which r indicates the number of the indices $ab \ldots c$ and g is the determinant of the matrix $(10 \cdot 1)$,

$$(10 \cdot 9) \qquad g = \epsilon^{pq \ldots r} g_{p1} g_{q2} \ldots g_{rn}$$

$$= \frac{1}{n!} \epsilon^{pq \ldots r} \epsilon^{ij \ldots m} g_{pi} g_{qj} \ldots g_{rm}.$$

As a special case of $(10 \cdot 8)$

$$(10 \cdot 10) \qquad g_{pj} G^{p;k} = g \delta^k_j.$$

If we define the quantities g^{ij} by the equation

$$(10 \cdot 11) \qquad g^{ij} = \frac{G^{i;j}}{g},$$

we have, therefore,

$$(10 \cdot 12) \qquad g_{ij} g^{ik} = \delta^k_j.$$

It is also evident that

$$(10 \cdot 13) \qquad \text{if } g_{ij} = g_{ji}, \text{ then } g^{ij} = g^{ji},$$

and that

$$(10 \cdot 14) \qquad |g^{ij}| = \frac{1}{g}.$$

The equation represented by $(10 \cdot 10)$ when $k = j$ leads by differentiation to

$$G^{i;j} = \frac{\partial g}{\partial g_{ij}}$$

because $G^{p;j}$ is independent of g_{pj}. This equation can be written

$$(10 \cdot 15) \qquad g^{ij} = \frac{1}{g} \frac{\partial g}{\partial g_{ij}}.$$

11. Historical.

Notations such as $sgn \binom{ij \ldots k}{ab \ldots c}$ for the sign of a permutation have long been used in algebra, but the idea of introducing 0 as a value of the symbol when the first row is not a permutation of the second row seems to have arisen in connection with the tensor analysis when Ricci (*Memorie della Società Italiana delle Scienze*, Ser. 3, vol. 12 (1899), p. 76) introduced the tensors $\sqrt{g}\,\epsilon_{ij\ldots k}$ and $(1/\sqrt{g})\,\epsilon^{ij\ldots k}$ (sistemi E) in which g is the determinant of a covariant tensor of the second order. Independently of this, in 1900 (*Annals of Math.* 2nd ser. vol. 1, p. 185), E. H. Moore introduced the symbol

$$\pm_{ij\ldots k},$$

which is equivalent to our ϵ, and applied it to the proof of theorems on determinants. The symbols ϵ were first given explicitly in the form which we are using in A. S. Eddington's *Espace, Temps, et Gravitation*, Partie théorique, Section IV, Paris, 1921, and were used to prove theorems on determinants. In 1922, J. Lipka (*Rendiconti della R. Acc. N. dei Lincei*, vol. 31, p. 242) proved formulas equivalent to (8·4), (10·7). I developed the theory of determinants somewhat in detail along the lines of § 3 above in my lectures in 1923, and during one of these lectures J. W. Alexander pointed out the advantages of bringing in the tensor $\delta^{ij\ldots k}_{ab\ldots c}$. This tensor was independently discovered by F. D. Murnaghan at about the same time, and his results were published in the Bulletin of the *Am. Math. Soc.* vol. 30 (1924), p. 293; vol. 31 (1925), p. 323 and *Am. Math. Monthly*, vol. 32 (1925), p. 233. I learn from Professor E. T. Bell that the generalized Kronecker deltas were also discovered independently by C. M. Cramlet.

The general ideas underlying all these questions of notation go back at least to Grassmann and have appeared in mathematical literature in various forms. For many purposes it is convenient to denote the "alternating part" of a set of quantities (i.e. the result of multiplying it by an appropriate Kronecker delta and an appropriate factorial) by enclosing it in square brackets. Thus

$$\frac{1}{m!}\,\delta^{ij\ldots k}_{pq\ldots r}\,T_{ij\ldots k}=[T_{pq\ldots r}],$$

if the number of subscripts is m. The analogous "symmetric part" can be denoted analogously by round brackets. In his book, *Der Ricci-Kalkül* (Berlin, 1924), J. A. Schouten encloses in square brackets the indices with respect to which the alternating part is taken. He denotes the simple Kronecker delta by A^i_j and thus on p. 31 comes to the generalized Kronecker delta defined as follows:

$$A^{ijk}_{abc}=A^{[i}_a\,A^j_b\,A^{k]}_c=\frac{1}{3!}\,\delta^{ijk}_{abc}.$$

CHAPTER II

DIFFERENTIAL INVARIANTS

1. N-dimensional space.

By a *space of n dimensions* we mean any set of objects which is in a one-to-one reciprocal correspondence with the totality of ordered sets of n real or n complex numbers $(x^1, x^2, ..., x^n)$ satisfying a set of inequalities

$$(1 \cdot 1) \qquad\qquad | x^i - A^i | < k^i$$

in which $A^1, A^2, ..., A^n$ are constants and $k^1, k^2, ..., k^n$ are positive constants. The objects themselves will habitually be called *points*, although in the applications of the theory they may be objects of very diverse sorts. For example, if $n = 1$ they may be the frequencies of the spectrum; if $n = 4$ they may be the events of space-time as pictured in the Einstein theory; for any value of n they may be the configurations of a dynamical system.

Any one-to-one reciprocal correspondence between the points and a set of ordered sets of numbers $(x^1, x^2, ..., x^n)$ will be called a *coordinate system*. The numbers, $x^1, x^2, ..., x^n$, are the coordinates of the point to which the set $(x^1, x^2, ..., x^n)$ corresponds in the coordinate system. We shall use the single letter x to stand for the ordered set of letters $(x^1, x^2, ..., x^n)$.

It is in general a physical rather than a mathematical problem to set up a coordinate system. For example, an important branch of astronomy is concerned with attaching coordinates to the stars. But it may also be a mathematical problem, as is the case when geometry is based on a set of postulates and a series of abstract constructions lead to a coordinate system.

2. Transformations of coordinates.

A set of n equations

$$(2 \cdot 1) \qquad\qquad \bar{x}^i = f^i (x^1, x^2, ..., x^n)$$

in which the functions f^i are single-valued for all points x of our n-dimensional space and which can be solved so as to yield a set of n equations

$$(2 \cdot 2) \qquad\qquad x^i = g^i (\bar{x}^1, \bar{x}^2, ..., \bar{x}^n),$$

in which the functions g^i are single-valued for all $\bar{x}$'s given by (2·1) and (1·1), determines a correspondence between the sets of coordinates x and the sets of numbers $\bar{x}$. If we regard each set of numbers $\bar{x}$ as coordinates of the same point as the corresponding x, we have a new way of attaching the numbers to the points, i.e. a new coordinate system. From this point of view (2·1) defines a *transformation of coordinates*.

It is also possible to look on these equations (2·1) as defining a transformation of the points. For x and $\bar{x}$ may be regarded as coordinates of different points in the same coordinate system, and then the equations determine a rule by which each point corresponds to another point. At present, however, we shall adopt the point of view of transformations of coordinates, according to which x and $\bar{x}$ are coordinates of the same point, and the equations serve as a sort of dictionary which enables us to pass from the way of naming the points determined by the first coordinate system to the way of naming them determined by the second coordinate system.

The result of applying two of our transformations of coordinates, one after the other, is a transformation of coordinates, and each transformation has an inverse. Hence the set of all admitted transformations of coordinates forms a *group*. There is thus a perfectly definite family of coordinate systems before us, namely the arbitrary one with which we start* and all those obtainable from it by transformations (2·1). Since the transformations of coordinates which we have admitted form a group, it makes no difference which coordinate system we start with so long as it is one of the family.

3. Invariants.

An object of any sort which is not changed by transformations of coordinates is called an *invariant*. For example, any point is an invariant, for a transformation changes the coordinates x into $\bar{x}$ but does nothing to the point itself. A point is an invariant which has a unique set of coordinates in each coordinate system.

Any set of points is an invariant. So is any point function. If such a function is represented by $A(x)$ in one coordinate system, it will

---

* The initial coordinate system, as we have said, may be given by physical conditions. How it is determined has no effect on the rest of our argument. In the language of postulate theory, our undefined elements are (1) a class of individuals called points and (2) a class of (1-1) correspondences, called coordinate systems, between the points and the ordered sets of n numbers; and our unproved propositions state that the transformations of coordinates form a group of a particular type.

be represented by $\bar{A}\,(\bar{x})$, where

(3·1) $\bar{A}\,(\bar{x}) = A\,(g^1\,(\bar{x}),\,g^2\,(\bar{x}),\,...,\,g^n\,(\bar{x})),$

in any coordinate system $\bar{x}$ related to x by (2·2). The point function is therefore represented by a function of n variables, $A\,(x)$, in each coordinate system x. We shall call the point function an *absolute scalar* and the function $A\,(x)$ the *component* of the scalar in the coordinate system x. Thus a scalar is an invariant which has a unique component in each coordinate system. It is not hard to imagine that there are invariants which have several components in each coordinate system, and we shall soon see that this is the case.

4. Differential invariants.

What has been said about transformations of coordinates and invariants remains unchanged if we restrict the group of transformations by imposing further conditions on the functions f^i. But as the group of transformations is made smaller, the class of invariants will become more extensive. If the condition imposed is merely that the functions f^i shall be continuous, then the invariant theory which results will be Analysis Situs. We shall, however, impose also the condition that the functions f^i shall be analytic* and that the functional determinant

(4·1) $\left|\dfrac{\partial \bar{x}}{\partial x}\right| = \dfrac{\partial\,(\bar{x}^1\,...\,\bar{x}^n)}{\partial\,(x^1\,...\,x^n)}$

shall be different from zero for all points. Under these circumstances it is a theorem† that the equations (2·1) can be solved so as to yield a set of equations (2·2) valid in a neighbourhood of any point. Our general definition of a transformation of coordinates requires in addition that (2·2) shall be a single-valued transformation for all points.

By reference to equations (5·3) of Chap. I it is evident that the resultant of two transformations of the restricted type is a transformation of the same type; and by reference to (5·5), Chap. I, that the inverse of any one of the restricted transformations of coordinates is of the same type. Hence the restricted transformations form a group. From now on we shall mean a transformation of this group

---

* From now on we shall use the word function in all connections to signify an analytic function.

† This and other theorems on functional determinants needed in our work will be found in Goursat's *Mathematical Analysis*, New York, 1904, Chap. II, § 188, vol. 1, and § 98, pt. 1, vol. 2.

whenever we refer to a transformation of coordinates and we shall mean one of a family of coordinate systems related by transformations of this group whenever we refer to a coordinate system. An invariant under this group is called a *differential invariant*. The adjective will be found to be appropriate both because the operations of the differential calculus often enter into the definition of these invariants and because the invariants are often functions of sets of differentials. The essential point about them, however, is that they are unchanged by the group of all transformations of coordinates of the sort we have just described.

5. Differentials and contravariant vectors.

Any transformation of coordinates is *locally linear*. For if e is an infinitesimal and dx^1, dx^2, ..., dx^n are variables not all zero, but all less than e, the transformation (2·1) changes the coordinates of two points $(x^1, x^2, ..., x^n)$ and $(x^1 + dx^1, x^2 + dx^2, ..., x^n + dx^n)$ into $(\bar{x}^1, \bar{x}^2, ..., \bar{x}^n)$ and $(\bar{x}^1 + d\bar{x}^1 + e^1, \bar{x}^2 + d\bar{x}^2 + e^2, ..., \bar{x}^n + d\bar{x}^n + e^n)$ respectively, where e^1, e^2, ..., e^n are infinitesimals of higher order than e and

$$(5·1) \qquad d\bar{x}^i = \frac{\partial \bar{x}^i}{\partial x^a} dx^a.$$

Thus with each set of values of the variables x there is associated a set of independent variables dx^1, dx^2, ..., dx^n which undergo the linear homogeneous transformation (5·1) when the x's undergo a transformation (2·1). The variables dx^1, dx^2, ..., dx^n are called the *differentials* of the variables x^1, x^2, ..., x^n, respectively, at the point x. The set of differentials will be denoted briefly by dx.

An arbitrary set of differentials in one coordinate system determines a unique set of differentials in any other coordinate system, and the two sets of differentials in any two coordinate systems are related by the formulas (5·1). The abstract object determined by this set of sets of differentials, together with the correspondence by which one set of differentials is determined for each coordinate system, is called an absolute *contravariant vector**. The differentials which it determines in any coordinate system are called its *components* in that coordinate system. It is an invariant because a

---

* The process of abstraction involved in this definition is essentially the same as that used in arriving at the concept of a class, or of a cardinal number, or of any of the other primordial terms used in mathematics. The relation between our conception of a vector and that generally accepted in elementary geometry and "vector analysis" will be discussed in Chap. IV.

transformation of coordinates changes one set of components into another according to the formula (5·1) but does nothing to the contravariant vector itself.

The totality of contravariant vectors dx associated with a point x constitutes a space of n dimensions. Hence with each point x of the original space we have associated an n-dimensional space of contravariant vectors*. Any transformation (2·1) of the coordinates of the points of the original space brings about or *induces* a linear homogeneous transformation of the components of the contravariant vectors of each of these associated spaces.

These linear transformations form a group because: (1) the resultant of $d\bar{x}^i = \dfrac{\partial \bar{x}^i}{\partial x^a}\, dx^a$ and $d\tilde{x}^i = \dfrac{\partial \tilde{x}^i}{\partial \bar{x}^a}\, d\bar{x}^a$ is $d\tilde{x}^i = \dfrac{\partial \tilde{x}^i}{\partial x^a}\, dx^a$, the coefficients of the three transformations being related by

$$\frac{\partial \tilde{x}^i}{\partial x^j} = \frac{\partial \tilde{x}^i}{\partial \bar{x}^a}\frac{\partial \bar{x}^a}{\partial x^j};$$

and (2) the inverse of (5·1) is

$$(5\cdot2) \qquad\qquad dx^i = \frac{\partial x^i}{\partial \bar{x}^a}\, d\bar{x}^a.$$

The group of linear homogeneous transformations thus determined at each point is isomorphic with the group of all transformations in such a way that each transformation (2·1) corresponds to a unique transformation (5·1) at any given point and each transformation (5·1) is determined by an infinity of transformations (2·1). This isomorphism enables us to apply the theorems of elementary algebra to the theory of transformations of coordinates and is one of the guiding principles of the theory of differential invariants.

6. A set of n functions $V^1(x)$, $V^2(x)$, ..., $V^n(x)$ determines a contravariant vector $dx^i = V^i(x)$ at each point where the functions are defined. The components of these vectors in any other coordinate system $\bar{x}$ are, according to (5·1), given by the functions $\bar{V}^1$, $\bar{V}^2$, ..., $\bar{V}^n$ such that

$$(6\cdot1) \qquad \bar{V}^i(\bar{x}) = V^1\frac{\partial \bar{x}^i}{\partial x^1} + V^2\frac{\partial \bar{x}^i}{\partial x^2} + \cdots + V^n\frac{\partial \bar{x}^i}{\partial x^n}.$$

The system of contravariant vectors, one at each point where the functions V are defined, is called an (absolute) *contravariant vector*

* The reader may visualize this for the case $n = 2$ by thinking of the original space as a portion of a surface and the space of contravariant vectors associated with any point x as the tangent plane at this point.

field, but we shall generally refer to it briefly as "a contravariant vector."

A contravariant vector field determines a set of differential equations

(6·2) $$\frac{dx^i}{ds} = V^i,$$

which are unaltered in form by transformations of coordinates. For if we multiply both members of (6·2) by $\partial \bar{x}^j/\partial x^i$ and sum with respect to i we obtain

(6·3) $$\frac{d\bar{x}^j}{ds} = \bar{V}^j.$$

Any solution of these differential equations is of the form

(6·4) $x^i = \phi^i(s)$, where $|\, s - s_0 \,| < k$,

s_0 being a constant and k a positive constant which may be chosen so small that distinct values of s give distinct points x. The set of points satisfying (6·4) is a simple curve*, and by a well-known existence theorem there is one and only one curve satisfying (6·2) through each point. Thus a contravariant vector field determines a *linear congruence*, i.e. a family of curves such that through each point there is one curve and only one. A curve is an invariant because it is a set of points and a linear congruence is an invariant because it is a set of curves. The linear congruence is defined by the differential equations

(6·5) $$\frac{dx^1}{V^1} = \frac{dx^2}{V^2} = \dots = \frac{dx^n}{V^n}$$

just as well as by (6·2).

At any point a contravariant vector determines a set of contravariant vectors whose components are proportional to its own. The abstract object corresponding to all these vectors is called a *direction* and is obviously a differential invariant. Two sets of differentials $dx^1, dx^2, \dots, dx^n$ and $\delta x^1, \delta x^2, \dots, \delta x^n$ at the same point determine the same direction if and only if

(6·6) $$\frac{dx^1}{\delta x^1} = \frac{dx^2}{\delta x^2} = \dots = \frac{dx^n}{\delta x^n}.$$

The differential equations (6·5) determine a unique direction at each

---

* In this definition the parameter s may be either real or complex, although in the one case the curve is a one-dimensional and in the other case a two-dimensional manifold. This ambiguous usage is desirable in discussions like the present one, which are mainly concerned with formal questions.

point where the functions V^i exist and are not all zero. A curve determines a unique direction at each of its points by means of the equations

$$(6\cdot7) \qquad dx^i = \frac{dx^i}{ds}\, ds.$$

The differentials are homogeneous coordinates of a direction and non-homogeneous coordinates of a contravariant vector.

7. A general class of invariants.

There is a very general class of invariants which is such that the invariant determines in any coordinate system a certain number of functions, $F_1\,(x, dx, \delta x\ ...)$, $F_2\,(x, dx, \delta x\ ...)$, ..., $F_M\,(x, dx, \delta x\ ...)$ of the coordinates and of sets of differentials. These functions are called the *components* of the invariant in the coordinate system and uniquely determine the invariant. The components $\bar{F}_1, \bar{F}_2, ..., \bar{F}_M$ in a coordinate system $\bar{x}$ are connected with the components in any coordinate system x by a definite law of transformation which determines the properties of the invariant.

An example of such an invariant is an absolute scalar, the law of transformation being $(3\cdot1)$. Another example is a contravariant vector field, the law of transformation being $(6\cdot1)$.

For a fixed point the components $F_1, F_2, ..., F_M$ undergo the transformations of a certain group G^1 as the coordinates are subjected to the group G of all admitted transformations. The group G is isomorphic with G^1; that is to say, each transformation of G determines one transformation of G^1 in such a way that the product of two transformations of G corresponds to the product of the two corresponding transformations of G^1. Inversely, each transformation of G^1 corresponds to a sub-group of transformations of G.

In case $F_1, F_2, ..., F_M$ are the components of a contravariant vector, the group G^1 is the group of linear homogeneous transformations of the differentials referred to in § 5. And in a large and important class of invariants (tensors) the group G^1 is simply isomorphic with this linear homogeneous group or with one of its subgroups.

8. Tensors.

An invariant of the kind described in the last section is called a *tensor field* if (1) its components in any coordinate system x are

2-2

functions of the coordinates x alone and can be denoted by

$$(8 \cdot 1) \qquad T^{a_1 a_2 \ldots a_k}_{b_1 b_2 \ldots b_m}$$

in which the indices take integer values from 1 to n, and (2) the law of transformation is of the form

$$(8 \cdot 2) \qquad \bar{T}^{i_1 i_2 \ldots i_k}_{j_1 j_2 \ldots j_m} = \left| \frac{\partial x}{\partial \bar{x}} \right|^N T^{a_1 a_2 \ldots a_k}_{b_1 b_2 \ldots b_m} \frac{\partial \bar{x}^{i_1}}{\partial x^{a_1}} \cdots \frac{\partial \bar{x}^{i_k}}{\partial x^{a_k}} \frac{\partial x^{b_1}}{\partial \bar{x}^{j_1}} \cdots \frac{\partial x^{b_m}}{\partial \bar{x}^{j_m}}.$$

This law of transformation is linear and homogeneous in the components of the tensor field with coefficients which are rational functions of the coefficients of the transformation (5·1) of the differentials at each point.

The number N is called the *weight* of the tensor field. If $N = 0$, the tensor field is said to be *absolute*. In general it is said to be *relative*. A relative tensor field of weight 1 is called a *tensor density*.

The tensor field is said to be *covariant of order m and contravariant of order k*. Its *total order* is $m + k$. If $m = 0$, the tensor field is said to be *contravariant*; if $k = 0$, to be *covariant*; and if neither k nor m is zero, to be *mixed*.

The indices i_1, i_2, ..., i_k are called contravariant indices and are always written as superscripts; the indices j_1, j_2, ..., j_m are called covariant indices and always written as subscripts. The order of the indices is important because it determines, as it were, the position of each component in the system of components. As a memory device, it is worth noticing that in (8·2) the free (i.e. unrepeated) indices are on the dashed variables.

At each point where the functions (8·1) are defined, the tensor field determines a set of n^{k+m} numbers which are the components of a *tensor* at the point in question. The distinction between a tensor field and a tensor is exactly the same as the distinction drawn in § 6, between a vector field and a vector. Just as in the case of a vector, we shall ignore this distinction in our terminology whenever no confusion is likely to result and shall refer to a tensor field as a tensor.

Let us now turn to some examples of tensors of special types.

9. Relative scalars.

If $F(x)$ is an arbitrary function defined for a coordinate system x, the equation

$$(9 \cdot 1) \qquad \bar{F}(\bar{x}) = \left| \frac{\partial x}{\partial \bar{x}} \right|^N F(g^1(\bar{x}), \ldots, g^n(\bar{x})) = \left| \frac{\partial x}{\partial \bar{x}} \right|^N F(x)$$

defines a function $\bar{F}(\bar{x})$ in any other coordinate system $\bar{x}$. Moreover, the formula (9·1) insures that if $\tilde{F}(\tilde{x})$ is defined analogously,

$$(9\cdot2)\qquad \tilde{F}(\tilde{x}) = \left|\frac{\partial x}{\partial \tilde{x}}\right|^{N} F(x) = \left|\frac{\partial x}{\partial \bar{x}}\right|^{N} \left|\frac{\partial \bar{x}}{\partial \tilde{x}}\right|^{N} F(x) = \left|\frac{\partial \bar{x}}{\partial \tilde{x}}\right|^{N} \bar{F}(\bar{x}).$$

Hence the formula (9·1) determines a unique function in each coordinate system and, as is shown by (9·2), any two of the functions are related by a formula analogous to (9·1). The abstract object determined by this class of functions, one for each coordinate system, is called a *relative scalar of weight N*. Each function is called the *component* of the scalar in the coordinate system for which it is defined. Thus a scalar is a tensor which has a single component in each coordinate system. A relative scalar of weight 0 reduces to an absolute scalar as already defined. A relative scalar of weight 1 is called a *scalar density*.

To illustrate, suppose we have a three-dimensional distribution of matter of density $F(x^1, x^2, x^3)$, the coordinates (x^1, x^2, x^3) being rectangular cartesian. The mass contained in a given volume is given by the integral

$$M = \iiint F(x^1, x^2, x^3)\, dx^1 dx^2 dx^3$$

extended over the volume in question. If we make an arbitrary transformation of coordinates (2·1), we find

$$M = \iiint F(g^1(\bar{x}), g^2(\bar{x}), g^3(\bar{x})) \left|\frac{\partial x}{\partial \bar{x}}\right| d\bar{x}^1 d\bar{x}^2 d\bar{x}^3$$

$$= \iiint \bar{F}(\bar{x})\, d\bar{x}^1 d\bar{x}^2 d\bar{x}^3.$$

Thus in the coordinate system $\bar{x}$ the density of this distribution of matter must be represented by a function $\bar{F}$ given by the formula (9·1) with $N = 1$.

10. Covariant vectors.

The partial derivatives of an absolute scalar $A(x)$ are n functions of x,

$$(10\cdot1)\qquad\qquad A_i = \frac{\partial A}{\partial x^i}.$$

By differentiating the equation (3·1), which says that A is an absolute scalar, we find

$$(10\cdot2)\qquad\qquad \frac{\partial \bar{A}}{\partial \bar{x}^i} = \frac{\partial A}{\partial x^a}\frac{\partial x^a}{\partial \bar{x}^i}.$$

Hence the n functions of $\bar{x}$ defined by the equations

$$\bar{A}_i(\bar{x}) = \frac{\partial \bar{A}}{\partial \bar{x}^i}$$

are related to the functions $A_i(x)$ by n equations

(10·3) $\qquad \bar{A}_i = A_1 \frac{\partial x^1}{\partial \bar{x}^i} + A_2 \frac{\partial x^2}{\partial \bar{x}^i} + \dots + A_n \frac{\partial x^n}{\partial \bar{x}^i}.$

Thus the scalar A, together with any coordinate system, determines an ordered set of n functions $(A_1, A_2, \dots, A_n)$ according to (10·1) and the sets of functions determined by any two coordinate systems are related by the equations (10·3). The invariant whose components in any coordinate system x are the n functions $\partial A/\partial x^1$, $\partial A/\partial x^2, \dots, \partial A/\partial x^n$ is called the *gradient* of the scalar A.

A gradient is a special case of a covariant tensor of the first order. For by the definition in § 8, an absolute covariant tensor of the first order is an invariant whose components in a coordinate system x are n functions $A_1(x), A_2(x), \dots, A_n(x)$ and which has n components $\bar{A}_1(\bar{x}), \bar{A}_2(\bar{x}), \dots, \bar{A}_n(\bar{x})$ in any other coordinate system $\bar{x}$, which are defined by the law of transformation (10·3).

Any tensor of the first order is called a *vector*. Thus an absolute covariant tensor of the first order (e.g. a gradient) is an absolute covariant vector. An absolute contravariant tensor of the first order is an absolute contravariant vector as defined in § 6, and a relative contravariant vector has a law of transformation of the type

$$\bar{A}^i = \left| \frac{\partial x}{\partial \bar{x}} \right|^N A^a \frac{\partial \bar{x}^i}{\partial x^a}.$$

The distinction between the covariant and contravariant methods of transformation is perhaps most easily remembered by keeping in mind that the components of a contravariant vector transform like differentials and the components of a covariant vector like the derivatives of a scalar.

11. Algebraic combinations of tensors.

Further tensors can be constructed from given tensors by means of the three operations described in the following theorems.

The sum of two tensors of the same kind is a tensor of this kind. By this statement we mean that if we add the corresponding components of two tensors of the same weight N, the same number m of covariant indices, and the same number k of contravariant indices, the set of quantities thus obtained will be the components of a tensor of the same sort. Let us illustrate the theorem by means of two mixed tensor densities of the second order. By hypothesis,

$$\bar{T}^i_j = \left| \frac{\partial x}{\partial \bar{x}} \right| T^a_b \frac{\partial \bar{x}^i}{\partial x^a} \frac{\partial x^b}{\partial \bar{x}^j},$$

and

$$\bar{U}^i_j = \left| \frac{\partial x}{\partial \bar{x}} \right| U^a_b \frac{\partial \bar{x}^i}{\partial x^a} \frac{\partial x^b}{\partial \bar{x}^j}.$$

Hence

$$\bar{T}^i_j + \bar{U}^i_j = \left| \frac{\partial x}{\partial \bar{x}} \right| (T^a_b + U^a_b) \frac{\partial \bar{x}^i}{\partial x^a} \frac{\partial x^b}{\partial \bar{x}^j}.$$

The same proof will obviously apply in the general case.

The product of two tensors is a tensor. By this we mean that if we multiply each component of a tensor

$$T^{ij\ldots k}_{ab\ldots c}$$

of weight N having k contravariant and m covariant indices by each component of a tensor

$$U^{lm\ldots q}_{de\ldots f}$$

of weight N' having k' contravariant and m' covariant indices, we obtain a set of functions

$$T^{ij\ldots k}_{ab\ldots c} U^{lm\ldots q}_{de\ldots f} = V^{ij\ldots q}_{ab\ldots f},$$

which are the components of a tensor of weight $N + N'$, contravariant order $k + k'$, and covariant order $m + m'$. The proof consists simply in multiplying together the equations which give the law of transformation for the components of T and U respectively.

For example, the product of m absolute covariant vectors,

$$A^{(1)}_i, A^{(2)}_i, \ldots, A^{(m)}_i,$$

k absolute contravariant vectors, $V^i_{(1)}, V^i_{(2)}, \ldots, V^i_{(k)}$, and a relative scalar S of weight N,

$$V^{i_1}_{(1)} V^{i_2}_{(2)} \ldots V^{i_k}_{(k)} A^{(1)}_{j_1} A^{(2)}_{j_2} \ldots A^{(m)}_{j_m} S,$$

is a mixed tensor of weight N, covariant order m, and contravariant order k.

The third algebraic operation is that of *contraction*, which can be applied to any mixed tensor. It consists in setting a certain covariant index equal to a certain contravariant index and performing the indicated summation. The resulting set of functions are components of a tensor of order two less than the order of the original tensor. The proof may be illustrated as follows:

$$\bar{B}^{i}_{jkl} = B^{a}_{bcd} \frac{\partial \bar{x}^i}{\partial x^a} \frac{\partial x^b}{\partial \bar{x}^j} \frac{\partial x^c}{\partial \bar{x}^k} \frac{\partial x^d}{\partial \bar{x}^l}.$$

Hence
$$\bar{B}^{i}_{jil} = B^{a}_{bcd} \frac{\partial x^b}{\partial \bar{x}^j} \frac{\partial x^d}{\partial \bar{x}^l} \delta^{c}_{a}$$

$$= B^{a}_{bad} \frac{\partial x^b}{\partial \bar{x}^j} \frac{\partial x^d}{\partial \bar{x}^l}.$$

Hence
$$R_{jl} = B^{i}_{jil}$$

is a tensor, if B^{i}_{jkl} is a tensor.

12. Another rule, sometimes called the *quotient law*, is the following: if $V_{i} \, T^{ij...k}_{ab...c}$ is a tensor of the type indicated by the indices for all vectors V_{i}, then $T^{ij...k}_{ab...c}$ is a tensor of the type indicated by the indices. By hypothesis, we have

$$\bar{V}_{i} \bar{T}^{ij...k}_{lm...p} = \bar{V}_{i} \frac{\partial \bar{\bar{x}}^i}{\partial x^a} T^{ab...c}_{de...f} \frac{\partial \bar{x}^j}{\partial x^b} \cdots \frac{\partial \bar{x}^k}{\partial x^c} \frac{\partial x^d}{\partial \bar{x}^l} \cdots \frac{\partial x^f}{\partial \bar{x}^p}.$$

Since these equations hold for arbitrary choices of the functions $\bar{V}_{i}$, it follows that

$$\bar{T}^{ij...k}_{lm...p} = T^{ab...c}_{de...f} \frac{\partial \bar{x}^i}{\partial x^a} \cdots \frac{\partial \bar{x}^k}{\partial x^c} \frac{\partial x^d}{\partial \bar{x}^l} \cdots \frac{\partial x^f}{\partial \bar{x}^p}.$$

Obviously an analogous theorem holds in which the covariant vectors V_{i} are replaced by contravariant vectors.

13. The commonness of tensors.

Consider n^{k+m} arbitrary functions of x,

$$T^{i_1 i_2...i_k}_{j_1 j_2...j_m},$$

assigning indices in an arbitrary fashion. Let N be an arbitrary whole number. Then define the functions

$$\bar{T}^{i_1 i_2...i_k}_{j_1 j_2...j_m}$$

in any other coordinate system by means of the equations (8·2). The resulting systems of n^{k+m} functions are components of a tensor because they are such that any two systems of functions, say in the $\bar{x}$ and $\tilde{x}$ systems of coordinates, are related by equations of the form (8·2). In other words, *the law of transformation* (8·2) *is transitive.*

The proof will be evident in general if we write it out for the case where $k = 1$ and $m = 2$. We have by definition

$$(13\cdot 1) \qquad \bar{T}^p_{qr} = \left| \frac{\partial x}{\partial \bar{x}} \right|^N T^a_{bc} \frac{\partial \bar{x}^p}{\partial x^a} \frac{\partial x^b}{\partial \bar{x}^q} \frac{\hat{c}}{\hat{o}} \frac{v^c}{\bar{x}^r},$$

$$\tilde{T}^p_{qr} = \left| \frac{\partial x}{\partial \tilde{x}} \right|^N T^i_{jk} \frac{\partial \tilde{x}^p}{\partial x^i} \frac{\partial x^j}{\partial \tilde{x}^q} \frac{\partial x^k}{\partial \tilde{x}^r}.$$

Solving the latter set of equations by multiplying by

$$\left| \frac{\partial \tilde{x}}{\partial x} \right|^N \frac{\partial x^a}{\partial \tilde{x}^p} \frac{\partial \tilde{x}^q}{\partial x^b} \frac{\partial \tilde{x}^r}{\partial x^c},$$

and performing the indicated summations, we obtain

$$T^a_{bc} = \left| \frac{\partial \tilde{x}}{\partial x} \right|^N \tilde{T}^p_{qr} \frac{\partial x^a}{\partial \tilde{x}^p} \frac{\partial \tilde{x}^q}{\partial x^b} \frac{\partial \tilde{x}^r}{\partial x^c}.$$

When this is substituted in (13·1) the result is

$$\bar{T}^p_{qr} = \left| \frac{\partial \tilde{x}}{\partial \bar{x}} \right|^N \tilde{T}^a_{bc} \frac{\partial \bar{x}^p}{\partial \tilde{x}^a} \frac{\partial \tilde{x}^b}{\partial \bar{x}^q} \frac{\partial \tilde{x}^c}{\partial \bar{x}^r},$$

as the theorem requires.

14. Numerical tensors.

Suppose we require of a mixed tensor of the second order that its components in one coordinate system shall be δ^i_j. On account of the equations

$$(14\cdot 1) \qquad \delta^i_j = \delta^a_b \frac{\partial \bar{x}^i}{\partial x^a} \frac{\partial x^b}{\partial \bar{x}^j}$$

it follows that this tensor *has the same components in all coordinate systems*. We express this by saying that the Kronecker delta is a mixed tensor of the second order.

In general, a tensor whose components are constants in all coordinate systems will be called a *numerical tensor*. All the ϵ's and δ's of Chap. I are numerical tensors.

By (3·41) of Chap. I,

$$(14\cdot 2) \qquad \epsilon^{i_1 i_2 \ldots i_n} = \left| \frac{\partial x}{\partial \bar{x}} \right| \epsilon^{a_1 a_2 \ldots a_n} \frac{\partial \bar{x}^{i_1}}{\partial x^{a_1}} \frac{\partial \bar{x}^{i_2}}{\partial x^{a_2}} \cdots \frac{\partial \bar{x}^{i_n}}{\partial x^{a_n}}.$$

Hence the numbers $\epsilon^{i_1 i_2 \ldots i_n}$ *are the components of a contravariant tensor density* which has the same components in all coordinate systems. In like manner, (3·42) of Chap. I gives

$$(14\cdot 3) \qquad \epsilon_{i_1 i_2 \ldots i_n} = \left| \frac{\partial x}{\partial \bar{x}} \right|^{-1} \epsilon_{a_1 a_2 \ldots a_n} \frac{\partial x^{a_1}}{\partial \bar{x}^{i_1}} \frac{\partial x^{a_2}}{\partial \bar{x}^{i_2}} \cdots \frac{\partial x^{a_n}}{\partial \bar{x}^{i_n}}.$$

It follows that $\epsilon_{i_1 i_2 \ldots i_n}$ *are the components of a covariant tensor of weight* -1 which has the same components in all coordinate systems.

Each of the generalized Kronecker deltas defined in Chap. I determines an absolute tensor whose components are the same in all coordinate systems. For by (8·4), Chap. I, any Kronecker delta is obtainable by multiplying together two epsilons, one a relative tensor of weight $+1$ and the other a relative tensor of weight -1, and contracting with respect to a certain number of indices. The result is an absolute tensor, by the second and third theorems of § 11.

15. Combinations of vectors.

As a simple application of the theorems of § 11 we may observe that if A_i is a covariant and B^i a contravariant vector, then $A_i B^j$ is a mixed tensor of the second order. On contraction we find that

$$(15·1) \qquad A_i B^i = A_1 B^1 + A_2 B^2 + \ldots + A_n B^n$$

is a scalar. It is called the *scalar product* or *inner product* of the two vectors.

If A_i and B_i are two covariant vectors, it follows from § 11 that

$$(15·2) \qquad C_{ij} = A_i B_j - A_j B_i = \delta_{ij}^{ab} A_a B_b$$

is a covariant tensor of the second order. It is called the *outer product* of the two vectors. The same term is sometimes applied if $n = 3$ to the contravariant vector density

$$(15·3) \qquad c^i = \epsilon^{ijk} A_j B_k = \tfrac{1}{2} \epsilon^{ijk} C_{jk}.$$

The non-zero components of c^i and C_{ij} are the same, namely,

$$A_2 B_3 - A_3 B_2, \ A_3 B_1 - A_1 B_3, \ A_1 B_2 - A_2 B_1.$$

These two invariants are also closely connected with the vector product of A and B, which we shall define in § 8, Chap. IV.

The outer product of k covariant vectors $A_i^{(1)}, A_i^{(2)}, \ldots A_i^{(k)}$ is defined analogously as

$$(15·4) \qquad A_{i_1 i_2 \ldots i_k} = \delta_{i_1 i_2 \ldots i_k}^{a_1 a_2 \ldots a_k} A_{a_1}^{(1)} A_{a_2}^{(2)} \ldots A_{a_k}^{(k)}.$$

This is a covariant tensor. If $k = n - 1$, there is also a contravariant vector density,

$$(15·5) \qquad a^i = \epsilon^{i a_1 a_2 \ldots a_k} A_{a_1}^{(1)} A_{a_2}^{(2)} \ldots A_{a_k}^{(k)} = \frac{1}{k!} \epsilon^{i a_1 a_2 \ldots a_k} A_{a_1 a_2 \ldots a_k},$$

which has the same non-zero components. Many other combinations of covariant vectors can be formed, and of course there is an analogous set of theorems about contravariant vectors.

The k directions determined by k sets of differentials,

$$
\begin{array}{cccc}
d_1x^1 & d_1x^2 & \dots & d_1x^n \\
d_2x^1 & d_2x^2 & \dots & d_2x^n \\
\vdots & \vdots & & \vdots \\
d_kx^1 & d_kx^2 & \dots & d_kx^n,
\end{array}
$$

are said to be independent if and only if the k sets of differentials are linearly independent, that is to say, providing the determinants

$$ dS^{i_1 i_2 \dots i_k} = \delta^{i_1 i_2 \dots i_k}_{a_1 a_2 \dots a_k} d_1 x^{a_1} d_2 x^{a_2} \dots d_k x^{a_k} $$

are not all zero. These determinants are the components of a contravariant tensor which appears in the multiple integral of an arbitrary covariant tensor

$$ \int \dots k \dots \int T_{pq \dots r} \, dS^{pq \dots r}. $$

In the literature this integral is usually defined only for alternating tensors. This restriction makes no essential difference because if A is the alternating part of T we have (cf. (2·3) and (2·4), Chap. I),

$$ k! \, A_{pq \dots r} = \delta^{ij \dots k}_{pq \dots r} T_{ij \dots k} $$

and $$ k! \, dS^{pq \dots r} = \delta^{pq \dots r}_{ij \dots k} dS^{ij \dots k}, $$

so that the integral above is equal to

$$ \int \dots k \dots \int A_{pq \dots r} \, dS^{pq \dots r}. $$

For the theory of multiple integrals the reader is referred to the fundamental memoir of H. Poincaré, *Acta Mathematica*, vol. 9 (1887), p. 321; to F. D. Murnaghan, *Vector analysis and the theory of relativity*, Baltimore, 1922; and to A. Buhl, *Intégrales Doubles*, Paris, 1920, and *Formules Stokiennes*, Paris, 1926.

16. Historical and general remarks.

The explicit recognition of any system of functions with a law of transformation as a definite object of study seems to be due to G. Ricci (*Atti della R. Acc. dei Lincei, Rendiconti*, Ser. 4, vol. 3, pt. 1 (1887), p. 15, or vol. 5, pt. 1 (1889), pp. 112 and 643). It is of course a natural generalization from the conceptions made familiar by the many systems of vector and geometrical analysis developed during the last century. Ricci also introduced the use of superscripts and subscripts (*Atti Lincei, Rendiconti*, vol. 5, pt. 1 (1889), pp. 112 and 643) to distinguish between the two particular laws of transformation known as covariant and contravariant and outlined all the essentials of the Tensor Analysis. The convenient though historically not well justified name, *tensor*, was introduced by A. Einstein, *Annalen der Physik*, 4th Series, vol. 49 (1916), p. 769, and popularized by the success of the theory of Relativity.

The importance of the theory of differential invariants in physical problems is due to the fact that when coordinates are used to describe physical phenomena (e.g. those studied in geometry) it is usually the case that the coordinates are no part of the phenomena themselves. They are generally put into the description by the observer. Therefore it is desirable to have the description in such terms that when stated in terms of one coordinate system, it can be read off easily in terms of other systems. Such statements will employ invariants of one sort or another.

The special importance of tensors is due to the fact that the components of a tensor in one coordinate system are linear homogeneous functions of its components in any other coordinate system. Hence, *if the components of a tensor are all zero in one coordinate system, the components vanish in all coordinate systems*. Hence, the vanishing of a tensor is the typical form of a physical or geometric theorem.

This matter of the physical significance of tensors is developed in a brilliant and entertaining way in Eddington's book on the *Mathematical Theory of Relativity*. Another good account of it is found in Einstein, *The Meaning of Relativity*. The reader is advised to read the account of tensor analysis in these or other books on relativity. Here we have space only for the purely mathematical point of view.

The various types of absolute tensors have been characterized by H. Weyl (*Math. Zeitschrift*, vol. 23 (1925), p. 271 and vol. 24, pp. 328 and 377) by the linear homogeneous character of their law of transformation and the isomorphisms of these transformations with the group of linear homogeneous transformations with determinant unity.

The invariant theory of systems of functions in the sense of Ricci underlies the modern differential geometry of infinitesimal displacements of various sorts, though this fact is not always clearly brought out. For a general account of this subject, with references, see

J. A. Schouten, *Rendiconti Circolo Mat. di Palermo*, vol. 50 (1926), p. 142. In this field, non-tensor invariants are becoming more and more important. Examples of such invariants are the affine connection (§ 10, Chap. III), the projective connection (T. Y. Thomas, *Proc. Nat. Ac. Sc.* vol. 11 (1925), p. 199), and the conformal connection (J. M. Thomas, *Proc. Nat. Ac. Sc.* vol. 11 (1925), p. 257). The terminology which these ideas have caused us to use diverges somewhat from that of the older writers on differential invariant theory, but it follows closely the usage indicated by the physical applications. And it seems to the writer that a theory which is chiefly valuable for its applications should enunciate its results in the form most suitable for such applications rather than in the form due to the accidents of its development.

CHAPTER III

QUADRATIC DIFFERENTIAL FORMS

1. Differential forms.

A scalar is an invariant of the type described in § 7, Chap. II, which has one component in each coordinate system, this component being a function of the coordinates alone. The second condition may be generalized by allowing the component to be also a function of a number of sets of differentials, while the law of transformation is, as before,

$$(1\cdot1) \qquad \bar{F}(\bar{x}, d\bar{x}, \delta\bar{x}, \ldots) = \left| \frac{\partial x}{\partial \bar{x}} \right|^N F(x, dx, \delta x, \ldots).$$

If the component is required to be a homogeneous polynomial in the differentials, the invariant is called a *relative differential form of weight N*. In case $N = 0$ the differential form is said to be *absolute*. We shall generally mean "absolute differential form" when we refer to a "differential form" without the adjective "relative."

2. Linear differential forms.

If the component of a differential form in any coordinate system is a linear homogeneous function $A_i dx^i$ of a single set of differentials with coefficients which are functions of x, the form is said to be *linear*. In this case the law of transformation

$$(2\cdot1) \qquad \bar{A}_i d\bar{x}^i = A_j dx^j = A_j \frac{\partial x^j}{\partial \bar{x}^i} d\bar{x}^i$$

shows that we must have

$$(2\cdot2) \qquad \bar{A}_i = A_j \frac{\partial x^j}{\partial \bar{x}^i}.$$

That is, *the coefficients of a linear differential form are the components of a covariant vector.*

Conversely, if $A_1, A_2, \ldots, A_n$ are the components of a covariant vector, $A_i dx^i$ is the component of a differential form. For if we multiply both sides of the equations (2·2) by $d\bar{x}^i$ and sum with respect to i we obtain (2·1). Hence the theory of linear differential forms is equivalent to that of absolute covariant vectors.

3. Quadratic differential forms.

If the component of a differential form in any coordinate system is a homogeneous quadratic function

$$(3 \cdot 1) \quad g_{ij} dx^i dx^j = g_{11} dx^1 dx^1 + g_{12} dx^1 dx^2 + \ldots + g_{1n} dx^1 dx^n$$
$$+ g_{21} dx^2 dx^1 + g_{22} dx^2 dx^2 + \ldots + g_{2n} dx^2 dx^n$$
$$\vdots$$
$$+ g_{n1} dx^n dx^1 + g_{n2} dx^n dx^2 + \ldots + g_{nn} dx^n dx^n,$$

with coefficients which are functions of x, the form is said to be *quadratic*. It is clear that no generality is lost by assuming that

$$(3 \cdot 2) \qquad\qquad g_{ij} = g_{ji}.$$

From the definition we have

$$\bar{g}_{ij} d\bar{x}^i d\bar{x}^j = g_{ab} dx^a dx^b = g_{ab} \frac{\partial x^a}{\partial \bar{x}^i} \frac{\partial x^b}{\partial \bar{x}^j} d\bar{x}^i d\bar{x}^j,$$

and therefore

$$(3 \cdot 3) \qquad\qquad \bar{g}_{ij} = g_{ab} \frac{\partial x^a}{\partial \bar{x}^i} \frac{\partial x^b}{\partial \bar{x}^j}.$$

Hence in every coordinate system the quadratic form determines a set of n^2 functions, $g_{ij}(x)$, and the sets of functions in any two coordinate systems are related by the law of transformation (3·3). Hence they are the components of a *covariant tensor of the second order*.

From (3·3) it follows that the relation (3·2) holds in all coordinate systems; hence the tensor g_{ij} is *symmetric*.

Just as in the case of linear differential forms it follows conversely that if g_{ij} is a symmetric covariant tensor of the second order $g_{ij} dx^i dx^j$ is a quadratic form. Hence the theory of quadratic differential forms is completely equivalent to that of symmetric covariant tensors.

4. What is meant by a cubic differential form, a quartic differential form, and so on, is now self-evident. The component of a *bilinear* differential form in any coordinate system x, is a function

$$A_{ij} dx^i \delta x^j$$

in which the coefficients A_{ij} are functions of x, and dx and δx are contravariant vectors at the point x. It follows without difficulty that the coefficients A_{ij} are the components of a covariant tensor of the second order.

The coefficients of the component of a differential form of arbitrary degree are the components of a covariant tensor. For each set of differentials which enters in more than the first degree these components can be assumed to be symmetric in the corresponding subscripts.

5. Invariants derived from basic invariants.

On account of the frequency with which differential invariants occur (cf. § 13, Chap. II) it is clear that the problem of the theory of differential invariants cannot be merely to find invariants. The real problem of the theory is to start with a given set of invariants—we shall call them the basic invariants—which arise in some physical or mathematical problem, and derive from them a system of new invariants which will serve the purposes of the problem in question. The operations generally employed in building new invariants are addition, subtraction, multiplication, division and differentiation. But irrational algebraic processes are also used in some cases.

The invariants thus constructed out of the basic invariants are called *invariants of* the basic ones. For example, the gradient $\partial A/\partial x^i$ is an invariant of a scalar A. In case there is more than one basic invariant, we sometimes speak of *simultaneous invariants* of the basic ones. For example, if A_i is a covariant and B^i a contravariant vector, the scalar product $A_i B^i$ is a simultaneous invariant of the two vectors.

6. Invariants of a quadratic differential form.

The determinant of the coefficients of a quadratic differential form $g_{ij} dx^i dx^j$ is a relative scalar of weight 2, since by (10·9), Chap. I, it is the result of contracting the product of n absolute tensors and two relative tensors each of weight 1. Hence $\sqrt{g}$ is a scalar density. Throughout the rest of the chapter we assume that

$$(6·1) \qquad\qquad g \neq 0.$$

In like manner, by (10·7), Chap. I, the cofactors $G^{i;j}$ of the elements g_{ij} in the determinant g are the components of a contravariant tensor of weight 2, which is an invariant of the quadratic differential form. It follows from other formulas of § 10, Chap. I, that the minors of any order of g are the components of an absolute tensor which is an invariant of the differential form: and the cofactors of any order are the components of a relative tensor of weight 2, which is an invariant of the differential form.

The functions g^{ij} defined by (10·11), Chap. I, are the components of an absolute contravariant tensor, because the numerator in (10·11) is a contravariant tensor of weight 2 and the denominator a scalar of weight 2. As proved in § 10, Chap. I, it is symmetric and satisfies the equations

$$(6\cdot2) \qquad g^{as}g_{ai} = \delta^s_i.$$

The tensors g_{ij} and g^{ij} are called respectively the *fundamental covariant* and *fundamental contravariant tensors* of the quadratic form $g_{ij}dx^i dx^j$.

7. If we set

$$(7\cdot1) \qquad u^i_j = \frac{\partial x^i}{\partial \bar{x}^j},$$

the equations (3·3) become

$$(7\cdot2) \qquad \bar{g}_{ij} = g_{ab}u^a_i u^b_j,$$

which is the law of transformation of the coefficients of the algebraic form in dx,

$$(7\cdot3) \qquad G_2 = g_{ij}dx^i dx^j,$$

when the variables dx undergo the linear homogeneous transformation

$$(7\cdot4) \qquad dx^i = u^i_a d\bar{x}^a.$$

The laws of transformation of the tensors which we found in § 6 are all obtained by combining the equations (7·2) with themselves by rational processes. The theory of these tensors therefore is a part of the algebraic invariant theory of the quadratic form (7·3)—and, on account of our limitations of space, we shall have to leave it with this remark.

8. The fundamental affine connection.

It is natural to inquire next about invariants, the components of which are functions of the first derivatives of the functions g_{ij} as well as of these functions themselves. On differentiating the law of transformation (3·3) we obtain

$$(8\cdot1) \qquad \frac{\partial \bar{g}_{ij}}{\partial \bar{x}^k} = \frac{\partial g_{ab}}{\partial x^c}\frac{\partial x^a}{\partial \bar{x}^i}\frac{\partial x^b}{\partial \bar{x}^j}\frac{\partial x^c}{\partial \bar{x}^k} + g_{ab}\frac{\partial^2 x^a}{\partial \bar{x}^i \partial \bar{x}^k}\frac{\partial x^b}{\partial \bar{x}^j} + g_{ab}\frac{\partial x^a}{\partial \bar{x}^i}\frac{\partial^2 x^b}{\partial \bar{x}^j \partial \bar{x}^k}.$$

By an inspection of these equations we can find that *there is no tensor whose components are rational functions of the first derivatives of g_{ij} or of these derivatives and the g_{ij}'s themselves.* For the law of transformation of such a tensor would be obtained by elimination of the second derivatives between (7·1), (7·2) and (8·1). But (8·1) can be transformed into a set of equations which give each second derivative explicitly in terms of the first derivatives. To do this we add together the two other equations which are obtained from (8·1) by cyclic permutations of the subscripts ijk, subtract (8·1) from the sum, and divide by 2. The result is

$$(8·2) \quad \frac{1}{2}\left(\frac{\partial \bar{g}_{ik}}{\partial \bar{x}^j} + \frac{\partial \bar{g}_{jk}}{\partial \bar{x}^i} - \frac{\partial \bar{g}_{ij}}{\partial \bar{x}^k}\right)$$

$$= \frac{1}{2}\left(\frac{\partial g_{ac}}{\partial x^b} + \frac{\partial g_{cb}}{\partial x^a} - \frac{\partial g_{ab}}{\partial x^c}\right)\frac{\partial x^a}{\partial \bar{x}^i}\frac{\partial x^b}{\partial \bar{x}^j}\frac{\partial x^c}{\partial \bar{x}^k} + g_{ab}\frac{\partial x^a}{\partial \bar{x}^k}\frac{\partial^2 x^b}{\partial \bar{x}^i \partial \bar{x}^j}.$$

On multiplying the two sides of this by the corresponding sides of

$$\bar{g}^{lk}\frac{\partial x^d}{\partial \bar{x}^l} = g^{dc}\frac{\partial \bar{x}^k}{\partial x^c},$$

which is equivalent to (3·3), we obtain

$$(8·3) \quad \frac{1}{2}\bar{g}^{lk}\left(\frac{\partial \bar{g}_{ik}}{\partial \bar{x}^j} + \frac{\partial \bar{g}_{jk}}{\partial \bar{x}^i} - \frac{\partial \bar{g}_{ij}}{\partial \bar{x}^k}\right)\frac{\partial x^d}{\partial \bar{x}^l}$$

$$= \frac{1}{2}g^{dc}\left(\frac{\partial g_{ac}}{\partial x^b} + \frac{\partial g_{bc}}{\partial x^a} - \frac{\partial g_{ab}}{\partial x^c}\right)\frac{\partial x^a}{\partial \bar{x}^i}\frac{\partial x^b}{\partial \bar{x}^j} + \frac{\partial^2 x^d}{\partial \bar{x}^i \partial \bar{x}^j},$$

which gives an explicit formula for each of the second derivatives in terms of the quantities u^i_j, g_{ij} and the first derivatives of g_{ij}. Moreover, the equations (8·1) can be derived from (8·3).

From a set of equations such as (7·1) and (7·2) which do not involve the second derivatives and (8·3) which give one and only one explicit expression for each second derivative it is obviously impossible to eliminate the second derivatives. Hence we have the theorem stated above in italics.

9. The combinations of the g's which appear in (8·2) are called *Christoffel 3-index symbols of the first kind.* We shall denote them by the following symbolism

$$(9·1) \quad \frac{1}{2}\left(\frac{\partial g_{ik}}{\partial x^j} + \frac{\partial g_{kj}}{\partial x^i} - \frac{\partial g_{ij}}{\partial x^k}\right) = \Gamma_{k;\,ij} = [k, ij].$$

The combinations of the g's which appear in (8·3) are called *Christoffel 3-index symbols of the second kind,* and we denote them by

$$(9·2) \quad \frac{1}{2}g^{ka}\left(\frac{\partial g_{ia}}{\partial x^j} + \frac{\partial g_{aj}}{\partial x^i} - \frac{\partial g_{ij}}{\partial x^a}\right) = \Gamma^k_{ij} = \{^k_{ij}\}.$$

The equation (8·3) is now written as

$$(9·3) \qquad \overline{\Gamma}^r_{pq} = \Gamma^k_{ij} \frac{\partial x^i}{\partial \overline{x}^p} \frac{\partial x^j}{\partial \overline{x}^q} \frac{\partial \overline{x}^r}{\partial x^k} + \frac{\partial^2 x^i}{\partial \overline{x}^p \partial \overline{x}^q} \frac{\partial \overline{x}^r}{\partial x^i}.$$

Since the functions Γ^i_{jk} are uniquely determined for each coordinate system and since there is a definite law of transformation (9·3) between the Γ's in any two coordinate systems they are the components of an invariant as defined in § 7, Chap. II. Because of its geometrical significance (cf. § 3, Chap. IV) this invariant is called the *fundamental affine connection* of the quadratic differential form $g_{ij} dx^i dx^j$.

By interchanging k and i in (9·1) and adding the resulting equation to (9·1) we obtain

$$(9·4) \qquad \frac{\partial g_{ik}}{\partial x^j} = [k, ij] + [i, kj] = g_{ka} \Gamma^a_{ij} + g_{ia} \Gamma^a_{kj}.$$

10. Affine connections in general.

The equations of transformation (9·3) have a transitive property: that is to say, a transformation from the coordinates $\overline{x}$ to $\tilde{x}$ is effected by the equations

$$(10·1) \qquad \tilde{\Gamma}^l_{ij} = \overline{\Gamma}^d_{ab} \frac{\partial \overline{x}^a}{\partial \tilde{x}^i} \cdot \frac{\partial \overline{x}^b}{\partial \tilde{x}^j} \frac{\partial \tilde{x}^l}{\partial \overline{x}^d} + \frac{\partial^2 \overline{x}^d}{\partial \tilde{x}^i \partial \tilde{x}^j} \frac{\partial \tilde{x}^l}{\partial \overline{x}^d},$$

and if we substitute (9·3) in this we find

$$(10·2) \qquad \tilde{\Gamma}^l_{ij} = \Gamma^d_{ab} \frac{\partial x^a}{\partial \tilde{x}^i} \frac{\partial x^b}{\partial \tilde{x}^j} \frac{\partial \tilde{x}^l}{\partial x^d} + \frac{\partial^2 x^d}{\partial \tilde{x}^i \partial \tilde{x}^j} \frac{\partial \tilde{x}^l}{\partial x^d},$$

so that the transformation from x to $\tilde{x}$ is of the same form as the transformations from x to $\overline{x}$ and from $\overline{x}$ to $\tilde{x}$.

Hence, if we start with a perfectly arbitrary set of functions Γ^i_{jk} in one coordinate system x and define the $\overline{\Gamma}$'s in any other coordinate system $\overline{x}$ by means of the equations (9·3), we have one set of functions Γ for each coordinate system and the Γ's for any two coordinate systems, x and $\overline{x}$, are connected by equations of the form (9·3). Hence the Γ's in any coordinate system are the components of an invariant. The invariant thus defined is called an *affine connection* and the Γ's themselves in any coordinate system are called the *components of affine connection* in this coordinate system.

In case the components of affine connection in one coordinate system satisfy the condition

$$(10·3) \qquad \Gamma^i_{jk} = \Gamma^i_{kj},$$

it is clear from the equations of transformation that this condition is satisfied in all coordinate systems. Such an affine connection is said to be *symmetric*. Some writers call a general affine connection a linear connection and reserve the word affine for the symmetric case. The components of a general affine connection can be written in the form

$$(10\cdot4) \qquad \Gamma^i_{jk} = \Lambda^i_{jk} + \Omega^i_{jk},$$

where

$$(10\cdot5) \qquad \Lambda^i_{jk} = \tfrac{1}{2}(\Gamma^i_{jk} + \Gamma^i_{kj}),$$

and

$$(10\cdot6) \qquad \Omega^i_{jk} = \tfrac{1}{2}(\Gamma^i_{jk} - \Gamma^i_{kj}).$$

By substituting the law of transformation (9·3) in (10·5) we find that the functions Λ^i_{jk} satisfy the same law of transformation as Γ^i_{jk} and hence are the components of an affine connection which is obviously symmetric. On substituting (9·3) in (10·6) we find that Ω^i_{jk} are the components of a mixed tensor of the third order with one contravariant and two covariant indices. This tensor is obviously alternating in its covariant indices. Hence we see that the invariant theory of a general affine connection reduces to the theory of simultaneous invariants of a symmetric affine connection and an alternating tensor. From now on, therefore, we shall deal only with symmetric affine connections. What we have to say at first will apply to arbitrary symmetric affine connections and not merely to those defined by the formulas (9·2).

11. Covariant differentiation.

A tensor which is a simultaneous invariant of a given tensor and a given affine connection can be found by eliminating the second derivatives which appear in the law of transformation of the components of affine connection and in the formulas which result from differentiating the law of transformation of the components of the given tensor. For example, if V^i is an arbitrary contravariant vector, on differentiating its law of transformation we find

$$(11\cdot1) \qquad \frac{\partial \overline{V}^r}{\partial \overline{x}^q} = \frac{\partial V^k}{\partial x^j}\frac{\partial \overline{x}^r}{\partial x^k}\frac{\partial x^j}{\partial \overline{x}^q} + V^k \frac{\partial^2 \overline{x}^r}{\partial x^k \partial x^j}\frac{\partial x^j}{\partial \overline{x}^q}.$$

If we multiply (9·3) by $\overline{V}^p$ and sum with respect to p the result is

$$(11\cdot2) \qquad \overline{V}^p \overline{\Gamma}^r_{pq} = V^i \Gamma^k_{ij}\frac{\partial \overline{x}^r}{\partial x^k}\frac{\partial x^j}{\partial \overline{x}^q} + V^i \frac{\partial \overline{x}^p}{\partial x^i}\frac{\partial^2 x^j}{\partial \overline{x}^p \partial \overline{x}^q}\frac{\partial \overline{x}^r}{\partial x^j}.$$

When we add (11·1) and (11·2) the terms involving second derivatives cancel out on account of (6·1), Chap. I. We are therefore left with the equations which state that

$$(11\text{·}3) \qquad V^k_{,j} = \frac{\partial V^k}{\partial x^j} + V^i\, \Gamma^k_{ij}$$

is a mixed tensor. This tensor is called the *covariant derivative* of V^i.

If V_i is an absolute covariant vector, we find on differentiating its law of transformation

$$(11\text{·}4) \qquad \frac{\partial \overline{V}_r}{\partial \overline{x}^q} = \frac{\partial V_j}{\partial x^k}\frac{\partial x^j}{\partial \overline{x}^r}\frac{\partial x^k}{\partial \overline{x}^q} + \overline{V}_p\frac{\partial \overline{x}^p}{\partial x^j}\frac{\partial^2 x^j}{\partial \overline{x}^q \partial \overline{x}^r}.$$

When equations (9·3) are multiplied on both sides by $\overline{V}_r$ and subtracted from (11·4) they give

$$\frac{\partial \overline{V}_r}{\partial \overline{x}^q} - \overline{V}_a\, \overline{\Gamma}^a_{rq} = \left(\frac{\partial V_j}{\partial x^k} - V_a\,\Gamma^a_{jk}\right)\frac{\partial x^j}{\partial \overline{x}^r}\frac{\partial x^k}{\partial \overline{x}^q},$$

equations which state that

$$(11\text{·}5) \qquad V_{i,j} = \frac{\partial V_i}{\partial x^j} - V_a\,\Gamma^a_{ij}$$

are the components of a covariant tensor of the second order. The tensor $V_{i,j}$ is called the *covariant derivative* of the vector V_i.

If F is a relative scalar of weight N,

$$\overline{F} = F\left|\frac{\partial x}{\partial \overline{x}}\right|^N.$$

Hence, using (7·2) of Chap. I,

$$(11\text{·}6) \qquad \frac{\partial \overline{F}}{\partial \overline{x}^i} = \frac{\partial F}{\partial x^j}\frac{\partial x^j}{\partial \overline{x}^i}\left|\frac{\partial x}{\partial \overline{x}}\right|^N + NF\frac{\partial^2 x^a}{\partial \overline{x}^i\partial \overline{x}^b}\frac{\partial \overline{x}^b}{\partial x^a}\left|\frac{\partial x}{\partial \overline{x}}\right|^N.$$

On setting $r = q$ in (9·3) and summing we obtain

$$(11\text{·}7) \qquad \overline{\Gamma}^r_{ir} = \Gamma^a_{ja}\frac{\partial x^j}{\partial \overline{x}^i} + \frac{\partial^2 x^a}{\partial \overline{x}^i\partial \overline{x}^b}\frac{\partial \overline{x}^b}{\partial x^a}.$$

Hence, on eliminating second derivatives between (11·6) and (11·7),

$$(11\text{·}8) \qquad \frac{\partial \overline{F}}{\partial \overline{x}^i} - N\overline{F}\,\overline{\Gamma}^a_{ia} = \left(\frac{\partial F}{\partial x^j} - NF\,\Gamma^a_{ja}\right)\frac{\partial x^j}{\partial \overline{x}^i}\left|\frac{\partial x}{\partial \overline{x}}\right|^N.$$

That is to say, $\qquad F_{,i} = \dfrac{\partial F}{\partial x^i} - NF\,\Gamma^a_{ia}$

is a relative vector of weight N. It is called the *covariant derivative* of F.

12. By the elimination processes used in these three examples, there is no difficulty in proving that if $T^{ij...k}_{lm...n}$ is a relative tensor of weight N, then

$$(12\cdot1) \quad \begin{aligned} T^{ij...k}_{lm...n,p} = & \frac{\partial T^{ij...k}_{lm...n}}{\partial x^p} - NT^{ij...k}_{lm...n}\Gamma^a_{pa} \\ & + T^{aj...k}_{lm...n}\Gamma^i_{pa} + ... + T^{ij...a}_{lm...n}\Gamma^k_{pa} \\ & - T^{ij...k}_{am...n}\Gamma^a_{lp} - ... - T^{ij...k}_{lm...a}\Gamma^a_{np} \end{aligned}$$

is a relative tensor of weight N, whose covariant order is one greater than that of $T^{ij...k}_{lm...n}$. It is called the *covariant derivative* of $T^{ij...k}_{lm...n}$.

The process of forming the covariant derivative is referred to as covariant differentiation. We shall always indicate it notationally by a comma before the index which it adds to the tensor. If the tensor T is absolute, $N = 0$, and the second term in (12·1) drops out. If T is a contravariant tensor, the terms in the third line drop out. If T is covariant, the terms in the second line drop out. For example, in case of a covariant tensor of the second order, we have

$$T_{ij,k} = \frac{\partial T_{ij}}{\partial x^k} - T_{aj}\Gamma^a_{ki} - T_{ia}\Gamma^a_{kj}.$$

The process of forming the covariant derivative of a tensor can obviously be repeated indefinitely. The covariant derivative of the covariant derivative of a tensor is called the second covariant derivative, and so on. By forming the first, second, third, ... covariant derivatives of a tensor T we obtain an infinite sequence of simultaneous invariants of T and the affine connection.

13. Geodesic coordinates.

The components of a tensor cannot all vanish at any point in any coordinate system unless they vanish at this point in all coordinate systems. But on account of the second derivatives which appear in its law of transformation, the same is not true of an affine connection. For consider the transformation defined by

$$(13\cdot1) \quad y^i = x^i - q^i + \tfrac{1}{2}(\Gamma^i_{ab})_q(x^a - q^a)(x^b - q^b) + ...,$$

in which the dots stand for arbitrary terms of higher order and the subscript q signifies that the quantity within the parentheses is evaluated at the point $x = q$. If we substitute (13·1) in the law of transformation of the components of affine connection (9·3) we find that the components Γ^{*i}_{jk} in the coordinate system y vanish at the point $y = 0$, that is to say,

$$(13\cdot2) \qquad (\Gamma^{*i}_{jk})_0 = 0.$$

In case the affine connection is the fundamental affine connection of a quadratic differential form, it follows from (9·4) that we have

$$(13\cdot3) \qquad \left(\frac{\partial g^*_{ij}}{\partial y^k}\right)_0 = 0$$

in the coordinate system y. Hence the fundamental quadratic differential form itself has the component

$$(13\cdot4) \quad g^*_{ij}\, dy^i dy^j = (g^*_{ij})_0\, dy^i dy^j + \frac{1}{2}\left(\frac{\partial^2 g^*_{ij}}{\partial y^a \partial y^b}\right)_0 dy^i dy^j y^a y^b + \cdots$$

in these coordinates. Any coordinates satisfying the condition (13·2) or the equivalent conditions (13·3) or (13·4) are called *geodesic coordinates*.

14. Formulas of covariant differentiation.

On account of (13·2) the components in geodesic coordinates of the covariant derivative of any tensor T when evaluated at the origin of geodesic coordinates are the same as the ordinary derivatives of the components of T in these coordinates at this point, i.e.

$$(14\cdot1) \qquad \left(\frac{\partial T^{*a\ldots b}_{i\ldots j}}{\partial y^k}\right)_0 = (T^{*a\ldots b}_{i\ldots j,k})_0.$$

Since the origin of geodesic coordinates can be chosen arbitrarily it follows that the covariant derivative of the sum or the product of two tensors can be formed by the same rules as hold for ordinary differentiation. That is to say, if

$$(14\cdot2) \qquad T^{p\ldots q}_{i\ldots j} = A^{p\ldots q}_{i\ldots j} + B^{p\ldots q}_{i\ldots j},$$

then

$$(14\cdot3) \qquad T^{p\ldots q}_{i\ldots j,k} = A^{p\ldots q}_{i\ldots j,k} + B^{p\ldots q}_{i\ldots j,k},$$

and if

$$(14\cdot4) \qquad T^{p\ldots qr\ldots s}_{i\ldots jk\ldots l} = A^{p\ldots q}_{i\ldots j} B^{r\ldots s}_{k\ldots l},$$

then

$$(14\cdot5) \qquad T^{p\ldots qr\ldots s}_{i\ldots jk\ldots l,m} = A^{p\ldots q}_{i\ldots j,m} B^{r\ldots s}_{k\ldots l} + A^{p\ldots q}_{i\ldots j} B^{r\ldots s}_{k\ldots l,m}.$$

Since the components of the numerical tensors (the ϵ's and δ's) of § 14, Chap. II, are all constant in geodesic coordinates as well as in any coordinate system, it follows that their covariant derivatives are identically zero.

15. The equations (9·4) are equivalent to

$$(15\cdot1) \qquad g_{ij,\,k} = 0.$$

Since (9·2) and (9·4) are equivalent to each other we can say that the *fundamental affine connection of a quadratic differential form is that affine connection with respect to which the covariant derivative of the fundamental covariant tensor is zero.*

If we look at the formulas in § 10, Chap. I, for the determinant g and its various minors and cofactors, we see that, when differentiated, every term of these expressions contains a factor $\partial g_{ij}/\partial x^k$. Hence, if we carry out the differentiation process in geodesic coordinates and use (13·3), we must find in every case that the covariant derivative is zero at the origin of geodesic coordinates. In particular we have

$$(15\cdot2) \qquad g_{,\,i} = 0,$$

and

$$(15\cdot3) \qquad g^{ij}_{,\,k} = 0$$

at the origin of geodesic coordinates. Since the left-hand member of each of these equations is the component of a tensor, the equations hold in all coordinates. Since the origin of geodesic coordinates can be chosen arbitrarily, (15·2) and (15·3) hold for all points.

When expanded (15·2) and (15·3) give

$$(15\cdot21) \qquad \frac{\partial g}{\partial x^i} - 2g\,\Gamma^a_{ia} = 0,$$

and

$$(15\cdot31) \qquad \frac{\partial g^{ij}}{\partial x^k} + g^{ia}\,\Gamma^j_{ak} + g^{aj}\,\Gamma^i_{ak} = 0,$$

respectively.

From the formula (10·15) of Chap. I for the derivative of a determinant we find

$$(15\cdot4) \qquad \frac{\partial g}{\partial x^k} = \frac{\partial g_{ij}}{\partial x^k}\,g^{ij}g.$$

Hence by (15·21)

$$(15\cdot5) \qquad \Gamma^i_{ik} = \frac{1}{2}\frac{\partial \log g}{\partial x^k} = \frac{1}{2}\,g^{ij}\,\frac{\partial g_{ij}}{\partial x^k} = -\frac{1}{2}\,g_{ij}\,\frac{\partial g^{ij}}{\partial x^k}.$$

The functions Γ^a_{ai} which appear in these formulas have the law of transformation (11·7) which may be written

(15·6) $$\bar{\Gamma}^a_{ai} = \Gamma^a_{ab}\frac{\partial x^b}{\partial \bar{x}^i} + \frac{\partial}{\partial \bar{x}^i}\log\left|\frac{\partial x}{\partial \bar{x}}\right|.$$

16. The curvature tensor.

The process of covariant differentiation is not, in general, commutative with itself. For example, if V^i is a contravariant vector, on differentiating

$$V^i_{,j} = \frac{\partial V^i}{\partial x^j} + V^a \Gamma^i_{aj}$$

covariantly we obtain

$$V^i_{,j,k} = \frac{\partial^2 V^i}{\partial x^j \partial x^k} + \frac{\partial V^a}{\partial x^k}\Gamma^i_{aj} + V^a\frac{\partial \Gamma^i_{aj}}{\partial x^k} + \frac{\partial V^a}{\partial x^j}\Gamma^i_{ak} - \frac{\partial V^i}{\partial x^a}\Gamma^a_{jk}$$
$$+ V^a\Gamma^b_{aj}\Gamma^i_{bk} - V^a\Gamma^i_{ab}\Gamma^b_{jk},$$

and hence

(16·1) $$V^i_{,j,k} - V^i_{,k,j} = V^a B^i_{ajk},$$

in which

(16·2) $$B^i_{ajk} = \frac{\partial \Gamma^i_{aj}}{\partial x^k} - \frac{\partial \Gamma^i_{ak}}{\partial x^j} + \Gamma^b_{aj}\Gamma^i_{bk} - \Gamma^b_{ak}\Gamma^i_{bj}.$$

By the quotient law (§ 12, Chap. II), (16·1) implies that B^i_{ajk} is a tensor. It is called the *curvature tensor*. By (16·2) it satisfies the conditions

(16·3) $$B^i_{ajk} + B^i_{akj} = 0,$$

and

(16·4) $$B^i_{ajk} + B^i_{jka} + B^i_{kaj} = 0.$$

The curvature tensor and its successive covariant derivatives

(16·5) $$B^i_{ajk,l}, B^i_{ajk,l,m}, \cdots$$

are an infinite sequence of invariants of the affine connection, and therefore of the basic quadratic differential form, if the affine connection is the one defined by (15·1).

By contraction the curvature tensor gives rise to two tensors of the second order. One of these is

(16·6) $$S_{st} = B^a_{ast} = \frac{\partial \Gamma^a_{as}}{\partial x^t} - \frac{\partial \Gamma^a_{at}}{\partial x^s},$$

which obviously satisfies the condition

$$(16 \cdot 7) \qquad\qquad S_{ij} = -\, S_{ji}.$$

If we substitute (15·5) in (16·6), we see that

$$(16 \cdot 8) \qquad\qquad S_{ij} = 0$$

in case Γ is the fundamental affine connection of a quadratic differential form. The other tensor which is obtained by contracting the curvature tensor is

$$(16 \cdot 9) \qquad R_{ij} = B_{iaj}^{a} = \frac{\partial \Gamma_{ia}^{a}}{\partial x^{j}} - \frac{\partial \Gamma_{ij}^{a}}{\partial x^{a}} + \Gamma_{ia}^{p}\,\Gamma_{pj}^{a} - \Gamma_{ij}^{p}\,\Gamma_{pa}^{a},$$

which is called the Ricci tensor. The two tensors are related by

$$(16 \cdot 10) \qquad\qquad R_{ij} - R_{ji} = S_{ij},$$

so that according to (16·8) *the Ricci tensor of a quadratic differential form is symmetric.* For a quadratic differential form we also have a scalar invariant

$$(16 \cdot 11) \qquad\qquad R = g^{ij}R_{ij},$$

which is called the *scalar curvature.*

17. In geodesic coordinates (16·2) yields

$$(17 \cdot 1) \qquad\qquad (B_{ajk}^{*i})_0 = \left(\frac{\partial \Gamma_{aj}^{*i}}{\partial y^{k}} - \frac{\partial \Gamma_{ak}^{*i}}{\partial y^{j}} \right)_0 .$$

Writing (12·1) in geodesic coordinates, differentiating, evaluating at the origin and using (14·1), we find

$$
\begin{aligned}
(T_{pq...r,s,t}^{*ij...k})_0 = {}& \left(\frac{\partial^2 T_{pq...r}^{*ij...k}}{\partial y^{s}\partial y^{t}} \right)_0 - N \left(T_{pq...r}^{*ij...k} \frac{\partial \Gamma_{sa}^{*a}}{\partial y^{t}} \right)_0 \\
& + \left(T_{pq...r}^{*aj...k} \frac{\partial \Gamma_{sa}^{*i}}{\partial y^{t}} + ... + T_{pq...r}^{*ij...a} \frac{\partial \Gamma_{sa}^{*k}}{\partial y^{t}} \right)_0 \\
& - \left(T_{aq...r}^{*ij...k} \frac{\partial \Gamma_{ps}^{*a}}{\partial y^{t}} ... + T_{pq...a}^{*ij...k} \frac{\partial \Gamma_{rs}^{*a}}{\partial y^{t}} \right)_0 .
\end{aligned}
$$

Interchanging s and t and subtracting the resulting equation from this one, we obtain the following relation at the origin of geodesic coordinates:

$$
\begin{aligned}
(17 \cdot 2) \quad T_{pq...r,s,t}^{ij...k} - T_{pq...r,t,s}^{ij...k} = {}& T_{pq...r}^{aj...k} B_{ast}^{i} + T_{pq...r}^{ia...k} B_{ast}^{j} + ... \\
& + T_{pq...r}^{ij...a} B_{ast}^{k} - T_{aq...r}^{ij...k} B_{pst}^{a} - T_{pa...r}^{ij...k} B_{qst}^{a} - ... \\
& - T_{pq...a}^{ij...k} B_{rst}^{a} - N\, T_{pq...r}^{ij...k} S_{st}.
\end{aligned}
$$

Since all the terms are components of tensors, this holds in all coordinates, and since the origin of geodesic coordinates is arbitrary, it holds at all points. The formula (17·2) is known as *the Ricci identity*. (16·1) is a special case of it.

Another important identity, known by the name of Bianchi, is

(17·3) $$B^i_{ajk,l} + B^i_{akl,j} + B^i_{alj,k} = 0,$$

which is derived by differentiating (16·2) in geodesic coordinates, evaluating at the origin, and adding the three equations which are obtained by permuting the indices jkl cyclically.

18. Riemann-Christoffel tensor.

The tensor,

(18·1) $$R_{ijkl} = g_{ia} B^a_{jkl},$$

is called the *covariant curvature tensor* or the *Riemann-Christoffel* tensor. If we differentiate (18·1) covariantly, we find

(18·2) $$R_{ijkl,m} = g_{ia,m} B^a_{jkl} + g_{ia} B^a_{jkl,m}$$
$$= g_{ia} B^a_{jkl,m}$$

because of (15·1). In like manner,

(18·3) $$R_{ijkl,m,p} = g_{ia} B^a_{jkl,m,p},$$

and so on. Thus we have an infinite sequence of covariant tensors which are invariants of the quadratic differential form.

If we differentiate the relation

$$g_{ia} \Gamma^a_{jk} = \frac{1}{2} \left(\frac{\partial g_{ij}}{\partial x^k} + \frac{\partial g_{ik}}{\partial x^j} - \frac{\partial g_{jk}}{\partial x^i} \right) = [i, jk],$$

we find

$$\frac{\partial g_{ia}}{\partial x^l} \Gamma^a_{jk} + g_{ia} \frac{\partial \Gamma^a_{jk}}{\partial x^l} = \frac{1}{2} \left(\frac{\partial^2 g_{ij}}{\partial x^k \partial x^l} + \frac{\partial^2 g_{ik}}{\partial x^j \partial x^l} - \frac{\partial^2 g_{jk}}{\partial x^i \partial x^l} \right) = \frac{\partial [i, jk]}{\partial x^l}.$$

Interchanging k and l, and subtracting the resulting equation from this one and using (16·2), we obtain

(18·4) $$R_{ijkl} = \frac{1}{2} \left(\frac{\partial^2 g_{ik}}{\partial x^j \partial x^l} - \frac{\partial^2 g_{il}}{\partial x^j \partial x^k} - \frac{\partial^2 g_{jk}}{\partial x^i \partial x^l} + \frac{\partial^2 g_{jl}}{\partial x^i \partial x^k} \right)$$
$$+ g_{ab} (\Gamma^a_{ik} \Gamma^b_{jl} - \Gamma^a_{il} \Gamma^b_{jk})$$
$$= \frac{1}{2} \frac{\partial^2 g_{ab}}{\partial x^c \partial x^d} \delta^{ac}_{ij} \delta^{bd}_{kl} + g_{ab} (\Gamma^a_{ik} \Gamma^b_{jl} - \Gamma^a_{il} \Gamma^b_{jk})$$
$$= \frac{\partial [i, jk]}{\partial x^l} - \frac{\partial [i, jl]}{\partial x^k} + \Gamma^a_{ik} [a, jl] - \Gamma^a_{il} [a, jk].$$

From the first equality in (18·4) it is evident that

(18·5) $$R_{ijkl} = - R_{jikl} = - R_{ijlk},$$
(18·6) $$R_{ijkl} = R_{klij};$$

and from (16·4) that

(18·7) $$R_{ijkl} + R_{iklj} + R_{iljk} = 0.$$

On account of these identities there are at most $n^2 (n^2 - 1)/12$ distinct functions R_{ijkl}.

19. Since the components of any covariant tensor are the coefficients of a multilinear differential form, we have as invariants of the quadratic differential form

(19·1) $$G_2 = g_{ij} dx^i dx^j,$$

the following sequence of multilinear differential forms

(19·2) $$G_4 = R_{ijkl} d_1 x^i d_2 x^j d_3 x^k d_4 x^l,$$
(19·3) $$G_5 = R_{ijkl,m} d_1 x^i d_2 x^j d_3 x^k d_4 x^l d_5 x^m,$$
$$\vdots$$

The identities (18·5) imply

(19·4) $$R_{ijkl} = \tfrac{1}{2} \delta_{ij}^{ab} R_{abkl} = \tfrac{1}{2} \delta_{kl}^{cd} R_{ijcd}.$$

Hence the quadrilinear form (19·2) can be written (cf. § 15, Chap. II)

$$G_4 = \tfrac{1}{4} R_{abcd} \, dS^{ab} \, dT^{cd}$$
$$= \tfrac{1}{4} R_{abcd} (d_1 x^a d_2 x^b - d_2 x^a d_1 x^b)(d_3 x^c d_4 x^d - d_4 x^c d_3 x^d).$$

In view of (18·6) we may also embody the laws of transformation and of symmetry of the Riemann-Christoffel tensor in the statement that its components are the coefficients of the quadratic form

(19·21) $$\tfrac{1}{4} R_{ijkl} \, dS^{ij} \, dS^{kl}.$$

20. Reduction theorems.

The process by which we have been building invariants of a quadratic differential form is in outline as follows: We first observe that an arbitrary transformation of coordinates brings about the transformation

(20·1) $$dx^i = u_a^i \, d\bar{x}^a$$

in which

(20·2) $$u_a^i = \frac{\partial x^i}{\partial \bar{x}^a}.$$

Hence any tensor invariant whose components are functions of g_{ij} must be an algebraic invariant of the algebraic form

(20·3) $$g_{ij}dx^i dx^j = \bar{g}_{ij}d\bar{x}^i d\bar{x}^j.$$

Its law of transformation is obtained by eliminating the variables dx and $d\bar{x}$ between (20·1) and (20·3). This is equivalent to combining the equations

(20·4) $$\bar{g}_{ij} = g_{ab}\, u_i^a\, u_j^b$$

with themselves.

The tensor invariants whose components are functions of $\partial g_{ij}/\partial x^k$ as well as of g_{ij} must have laws of transformation obtained by eliminating between (20·4) and the law of transformation of the partial derivatives of the functions g_{ij}. This law, as proved in § 8, is equivalent to the law of transformation of the components of affine connection,

(20·5) $$\frac{\partial u_j^i}{\partial \bar{x}^k} = \bar{\Gamma}_{jk}^p\, u_p^i - \Gamma_{pq}^i\, u_j^p\, u_k^q\,.$$

As proved in § 8, there are no tensors whose components are rational functions of $\partial g_{ij}/\partial x^k$ as well as of g_{ij}. There is only the tensor equation

(20.6) $$g_{ij,\,k} = 0$$

which is equivalent to (20·5).

The tensor invariants whose components are rational functions of the second derivatives of g_{ij}, as well as of the first derivatives and of g_{ij}, have laws of transformation which result from eliminating

$$\frac{\partial u_j^i}{\partial \bar{x}^k} \text{ and } \frac{\partial^2 u_j^i}{\partial \bar{x}^k \partial \bar{x}^l}$$

between (20·4), (20·5), and the law of transformation of the second derivatives of g_{ij}. This law of transformation is obtained by differentiating (20·5) with respect to x^l. If we interchange k and l and subtract the equation thus obtained from the result of the differentiation, we get equations from which the second derivatives of u_j^i are eliminated. If we then substitute in these equations the values of the first derivatives of u_j^i from (20·5), the result is merely

(20·7) $$\bar{B}_{jkl}^p\, u_p^i = B_{qrs}^i\, u_j^q\, u_k^r\, u_l^s,$$

the law of transformation of the curvature tensor. Any further tensors are found by combining (20·4) and (20·7) or, what amounts to the same thing, (20·1), (20·3) and

$$(20\cdot 8) \qquad \bar{R}_{ijkl}d_1\bar{x}^i d_2\bar{x}^j d_3\bar{x}^k d_4\bar{x}^l = R_{ijkl}d_1 x^i d_2 x^j d_3 x^k d_4 x^l$$

in which it is understood that all the differentials undergo the same transformation (20·1).

To find the tensor invariants whose components are rational functions of the third derivatives of g_{ij} as well as of the lower derivatives and of the functions g_{ij} themselves we have to consider the law of transformation of the third derivatives of g_{ij}. This is obtained by differentiating (20·7). The derivatives of u^i_j are eliminated by substitution of (20·5), and the result is simply the law of transformation of the covariant derivative of the curvature tensor. That is to say, we are led to the invariants the laws of transformation of which result from the elimination of dx and $d\bar{x}$ from (20·1) and the equations

$$G_2 = \bar{G}_2, \;\; G_4 = \bar{G}_4, \;\; G_5 = \bar{G}_5.$$

By a continuation of this process we are led to the following theorem which may be said to reduce the problem of constructing differential invariants to one of algebraic invariant theory: *All tensor invariants of a quadratic differential form whose components are rational functions of the coefficients of the form and their first m derivatives are invariants of the algebraic forms G_2, G_4, ..., G_{m+2} under the linear transformation* (20·1), *and are rational functions of the coefficients of these forms.* The proof of this theorem will be made in Chap. VI (§ 8) by means of an extension of the idea underlying the geodesic coordinates.

The law of transformation of the covariant derivative of any tensor was seen in § 12 to be the result of eliminating second derivatives between the laws of transformation of the components of affine connection and of the derivatives of the components of the tensor. This leads to the following theorem, the proof of which is also postponed to Chap. VI:

If the components of a tensor are rational functions of the components of an affine connection and a set of basic tensors and their partial derivatives of the first m orders, they are rational functions of the components of the basic invariants, the curvature tensor, and their first m covariant derivatives. As a corollary we have the theorem: *If a basic set of tensors includes a symmetric tensor g^{ij} of the second order, all the tensor invariants of the basic set whose components are rational functions of the components of the basic invariants and their partial derivatives*

of the first m orders are rational functions of the coefficients of
G_2, G_4, ..., G_{m+2}, *the components of the given tensors, and their first m
covariant derivatives with respect to G_2.*

21. Historical remarks.

The first systematic study of a quadratic differential form in two
variables was the work of Gauss in 1827, *Disquisitiones generales
circa superficies curvas* (English translation, Princeton, 1902), in
which the scalar invariant called the curvature appears. The general
theory of the quadratic differential form was inaugurated by B.
Riemann in 1854 in his *Habilitationsschrift*, "Uber die Hypothesen
welche der Geometrie zu Grunde liegen" (*Werke*, 2nd ed. p. 272) and
developed more in detail in his "Commentatio mathematica, qua
respondere tentatur quaestioni ab Illma Accademia Parisiensi
propositae" (*Werke*, 2nd ed. p. 391). Before the details of Riemann's
work became known, the main lines of the theory were developed
by Lipschitz (cf. Chap. vi, § 18, below) and Christoffel (cf. Chap. v,
§ 13, below) both of whom found the components of affine connection
and the curvature tensor. Christoffel also used the operation of
covariant differentiation in order to derive the sequence of differential
forms G_4, G_5, etc. The name, covariant differentiation, and the
discovery of the general importance of this operation are due to
G. Ricci (*Atti della R. Acc. dei Lincei, Rendiconti*, Ser. 4, vol. 3,
pt. 1 (1887), p. 15). His studies and those of T. Levi-Civita were
set forth in their joint paper, "Méthodes de calcul différentiel absolu
et leurs applications," *Math. Ann.* vol. 54 (1900), p. 125.

The important rôle of the invariant whose components are the
Christoffel symbols of the second kind is largely due to the series of
geometrical investigations inaugurated by the paper of Levi-Civita
(*Rendiconti Circolo Mat. di Palermo*, vol. 42 (1917), p. 173), in which
he gave its geometrical interpretation by means of the concept of
infinitesimal parallelism. Components of affine connection which
are arbitrary functions symmetric in the subscripts were introduced
by H. Weyl ("Reine Infinitesimalgeometrie," *Math. Zeitschrift*, vol. 2,
(1918), p. 384) and given a geometric interpretation. For an account
of other researches in this direction by G. Hessenberg, J. A. Schouten,
R. König, W. Wirtinger, and E. Cartan, the reader may consult the
paper by Schouten referred to in Chap. ii, § 16, above, and for more
recent work the forthcoming Colloquium Lectures of Eisenhart on
Non-Riemannian Geometry, New York, 1927.

22. Scalar invariants.

The question naturally arises whether it is not possible to eliminate the variables u_j^i in (20·4) and in the laws of transformations of the coefficients of G_4, G_5, etc.

When this elimination can be carried out it leads to a number of equations

$$(22·1) \qquad I^{(a)}(R) = I^{(a)}(\bar{R})$$

in which R denotes the components of g_{ij}, R_{ijkl}, $R_{ijkl,m}$, and so on, and $\bar{R}$ the corresponding components of $\bar{g}_{ij}$, $\bar{R}_{ijkl}$, $\bar{R}_{ijkl,m}$, and so on. The functions I are absolute invariants in the sense of algebraic invariant theory. When the expressions for the quantities R as functions of x are substituted in these invariants we obtain a number of absolute scalars,

$$(22·2) \qquad S^{(a)}(x) = I^{(a)}(R),$$

which are invariants of the fundamental quadratic differential form.

The determination of the invariants I is a problem of algebraic invariant theory which we cannot treat here in detail. The algebraic theorem[*] is as follows: Let the laws of transformation of the coefficients of G_2, G_4, ..., G_N be written in the form

$$(22·3) \qquad \bar{R}^\beta = \phi^\beta(R, u).$$

Let M be the number of the functions ϕ which are independent as functions of the n^2 variables u_j^i. If $M > n^2$ the u's can be eliminated from (22·3) and there exist $M - n^2$ absolute invariants. These invariants are a *complete* set of absolute invariants in the sense that any other absolute invariant of the forms G_2, G_4, ..., G_N is a function of the I's.

The actual determination of a complete set of invariants in case $n = 3$ has been carried out by Christoffel in his 1869 paper referred to above.

If G_2 is a perfectly general quadratic differential form, a complete set of absolute invariants can be obtained from G_2 and G_4. But for particular G_2's it is necessary to go further out in the sequence of forms G_4, G_5, ... In certain cases, as in the Euclidean case treated in the next chapter, there is no value N such that the equations (22·3) determine the u's. We shall return to this subject at the end of Chap. v.

---

[*] On this question, see Capelli, *Lezioni sulla teoria delle forme algebriche*, p. 82 (Naples, 1902), and Weitzenböck, *Invariantentheorie*, p. 199 (Groningen, 1923).

A symbolic method of treating these differential invariants which is analogous to the symbolic method in algebraic invariant theory has been developed by H. Maschke and his students. For references, see Chap. v, § 3, below.

The problem of determining the scalar invariants of a quadratic form and any number of scalars A, B, C has also been attacked by a number of mathematicians from the point of view of the Lie theory of infinitesimal transformations. They inquire about functions of the variables $x, g_{ij}, \partial g_{ij}/\partial x^k, \partial^2 g_{ij}/\partial x^k \partial x^l, ..., A, B, ..., \partial A/\partial x^i, ...,$ etc., which are unaltered by infinitesimal transformations. This question gives rise to certain partial differential equations which must be satisfied by the invariants sought. From the properties of these differential equations the number of functionally independent scalar invariants of the basic invariants, which involve the components of these invariants and their derivatives up to an arbitrary fixed order, has been determined in some cases and some of the properties of these scalars have been worked out. Among the papers on this subject (an adequate account of which would require a. treatise at least as large as this Tract) are S. Lie, *Math. Ann.* vol. 24 (1884), p. 537; K. Zorawski, *Act. Math.* vol. 16 (1892), p. 1; T. Levi-Civita, *Atti d. R. Ist. Veneto*, Ser. 7, vol. 52 (1894), p. 1498; C. N. Haskins, *Trans. Amer. Math. Soc.* vol. 3 (1902), p. 71; vol. 5 (1904), p. 167; and vol. 7 (1906), p. 152; A. R. Forsyth, *Phil. Trans.* Ser. A, vol. 201 (1903), p. 329; vol. 202 (1903), p. 277; J. E. Wright, *Amer. Journ. of Math.* vol. 27 (1905), p. 323. The scalar invariants of affine connections are studied by T. Y. Thomas and A. D. Michal in the *Annals of Math.* vol. 28 (1927), p. 196; this memoir is to be followed by another one on relative quadratic forms.

CHAPTER IV

EUCLIDEAN GEOMETRY

1. Euclidean geometry.

In Euclidean geometry, in rectangular cartesian coordinates (y^1, y^2, y^3), the length of a curve joining two points A, B is defined by the formula*

$$(1\cdot1) \qquad \int_A^B ds = \int_{t_0}^{t_1} \sqrt{\left(\frac{dy^1}{dt}\right)^2 + \left(\frac{dy^2}{dt}\right)^2 + \left(\frac{dy^3}{dt}\right)^2} \, dt,$$

if the curve is given parametrically by equations

$$y^i = f^i(t),$$

in which A corresponds to $t = t_0$ and B to $t = t_1$. In arbitrary coordinates this formula becomes

$$(1\cdot2) \qquad \int_A^B ds = \int_{t_0}^{t_1} \sqrt{g_{ij} \frac{dx^i}{dt} \frac{dx^j}{dt}} \, dt,$$

in which

$$(1\cdot3) \qquad g_{ij} = \frac{\partial y^1}{\partial x^i} \frac{\partial y^1}{\partial x^j} + \frac{\partial y^2}{\partial x^i} \frac{\partial y^2}{\partial x^j} + \frac{\partial y^3}{\partial x^i} \frac{\partial y^3}{\partial x^j}.$$

If we transform to another coordinate system $\bar{x}$, the quantities g_{ij} in (1·2) are replaced by

$$\bar{g}_{ij} = \frac{\partial y^a}{\partial \bar{x}^i} \frac{\partial y^a}{\partial \bar{x}^j} = \frac{\partial y^a}{\partial x^p} \frac{\partial y^a}{\partial x^q} \frac{\partial x^p}{\partial \bar{x}^i} \frac{\partial x^q}{\partial \bar{x}^j} = g_{pq} \frac{\partial x^p}{\partial \bar{x}^i} \frac{\partial x^q}{\partial \bar{x}^j}.$$

Hence these quantities are the components of a covariant tensor of the second order, i.e., the coefficients of the quadratic differential form

$$(1\cdot4) \qquad ds^2 = g_{ij} dx^i dx^j = (dy^1)^2 + (dy^2)^2 + (dy^3)^2.$$

For example, if we make the transformation from rectangular cartesian to polar coordinates we find that

$$ds^2 = dr^2 + r^2 d\theta^2 + r^2 \cos^2 \theta \, d\phi^2$$

and the length of the curve is

$$\int_{t_0}^{t_1} \sqrt{\left(\frac{dr}{dt}\right)^2 + r^2 \left(\frac{d\theta}{dt}\right)^2 + r^2 \cos^2 \theta \left(\frac{d\phi}{dt}\right)^2} \, dt.$$

---

* We use the radical sign to indicate the positive value of the square root, or the principal value if we are dealing with complex quantities.

These formulas are all with respect to a fixed unit of length. It will be found that all the formulas of Euclidean geometry in which this unit of length is used can be written as equations between invariants of (1·4) or simultaneous invariants of (1·4) and other invariants. For this reason (1·4) is called the *fundamental quadratic differential form* of the Euclidean geometry and the tensors g_{ij} and g^{ij} (cf. § 6, Chap. III) are called respectively the *fundamental covariant* and *fundamental contravariant* tensor of the Euclidean geometry. In any rectangular cartesian coordinate system which employs the given unit of measure these tensors have the same components, namely

(1·5)
$$\begin{matrix} 1 & 0 & 0 \\ 0 & 1 & 0 \\ 0 & 0 & 1 \end{matrix},$$

but in polar coordinates their components are

$$\begin{matrix} 1 & 0 & 0 \\ 0 & r^2 & 0 \\ 0 & 0 & r^2\cos^2\theta \end{matrix} \quad \text{and} \quad \begin{matrix} 1 & 0 & 0 \\ 0 & \dfrac{1}{r^2} & 0 \\ 0 & 0 & \dfrac{1}{r^2\cos^2\theta} \end{matrix}$$

respectively. These two tensors and the Kronecker delta may be defined according to § 13, Chap. II, as (1) the covariant, (2) the contravariant, and (3) the mixed, tensor of the second order whose components are (1·5) in a particular coordinate system.

The three tensors are obviously related by the equations

(1·6)
$$g_{ia}g^{aj} = \delta_i^j$$

in the coordinate system in which they all have the components (1·5). Since both sides of (1·6) are tensors, (1·6) holds in all coordinate systems. It is in fact a special case of (6·2), Chap. III.

2. It is just as easy to write the formulas for space of an arbitrary number of dimensions as for $n = 3$. So we shall consider any quadratic differential form which reduces in a particular coordinate system y to

(2·1) $$ds^2 = (dy^1)^2 + (dy^2)^2 + \dots + (dy^n)^2$$

and refer to the theory of this differential form as *Euclidean geometry of n dimensions* with respect to a fixed unit of distance.

Any coordinate system y in which the quadratic form is given by the formula (2·1) will be called an *orthogonal coordinate system**, and it is obvious that the totality of such coordinate systems is an invariant of the quadratic form. Likewise any coordinate system y in which the g's are constants is called a *cartesian coordinate system*. The totality of these coordinate systems is, of course, also an invariant of the quadratic form. It includes the orthogonal coordinates as a sub-class, namely, the sub-class of those coordinate systems for which

$$(2\cdot2) \qquad\qquad g_{ij} = \delta^i_j.$$

The components of the *fundamental affine connection* of the differential form (2·1) and of the Euclidean geometry,

$$(2\cdot3) \qquad \Gamma^i_{jk} = \frac{1}{2} g^{ia} \left(\frac{\partial g_{ak}}{\partial x^j} + \frac{\partial g_{ja}}{\partial x^k} - \frac{\partial g_{jk}}{\partial x^a} \right),$$

are all zero in any cartesian coordinate system because the g's are constants. Conversely, if the Γ's are zero the condition (15·1), Chap. III, gives

$$g_{ij} = \text{constant}.$$

The law of transformation of the Γ's (cf. (9·3), Chap. III) now gives the following formula for the Γ's in any coordinate system x in terms of the transformation from x to the cartesian coordinate system y,

$$(2\cdot4) \qquad \Gamma^i_{jk} = \frac{\partial^2 y^a}{\partial x^j \partial x^k} \frac{\partial x^i}{\partial y^a},$$

or the equivalent formula

$$(2\cdot5) \qquad \Gamma^i_{jk} \frac{\partial y^p}{\partial x^i} = \frac{\partial^2 y^p}{\partial x^j \partial x^k}.$$

In case the coordinate systems x and y are both cartesian (2·5) reduces to

$$(2\cdot6) \qquad \frac{\partial^2 y^p}{\partial x^j \partial x^k} = 0,$$

which means that the y's are linear functions of the x's. In other words the transformation between any two cartesian coordinate systems is linear with constant coefficients,

$$(2\cdot7) \qquad\qquad y^i = p^i_a x^a + p^i.$$

---

* The term orthogonal coordinate system is generally used in a much broader sense. We are using it here in the restricted sense in order to avoid circumlocution.

Since the set of all transformations (2·7) for which the determinant $|p_j^i|$ is not zero obviously form a group it follows that the family of all cartesian coordinate systems is fully determined by any one of them.

For a transformation between two orthogonal coordinate systems y and $\bar{y}$, the law of transformation of the tensor g_{ij} (cf. (1·3)) reduces to

(2·8)
$$\delta_j^i = \frac{\partial y^a}{\partial \bar{y}^i} \frac{\partial y^a}{\partial \bar{y}^j}.$$

If we multiply both sides by $\partial \bar{y}^j / \partial y^b$ and sum with respect to j, these equations become

(2·9)
$$\frac{\partial \bar{y}^i}{\partial y^b} = \frac{\partial y^b}{\partial \bar{y}^i},$$

which are equivalent to

$$\delta_j^i = \frac{\partial y^i}{\partial \bar{y}^a} \frac{\partial y^j}{\partial \bar{y}^a}.$$

From (2·8) and the law of multiplication of determinants we have

(2·10)
$$\left| \frac{\partial y}{\partial \bar{y}} \right|^2 = 1, \quad \text{or} \quad \left| \frac{\partial y}{\partial \bar{y}} \right| = \pm 1.$$

Hence any transformation between orthogonal coordinate systems is of the form (2·7), in which the p's are constants satisfying the conditions

(2·11)
$$p_b^i p_b^j = p_i^a p_j^a = \delta_j^i.$$

Any matrix $\|p_j^i\|$ or transformation of coordinates (2·7) which satisfies these relations is said to be *orthogonal*. Thus a transformation between orthogonal coordinate systems is orthogonal. It is obvious from (2·9) that the resultant of any two orthogonal transformations is orthogonal; also that the inverse of any orthogonal transformation is orthogonal; and that the identity is an orthogonal transformation. Hence the orthogonal transformations form a group and any orthogonal coordinate system uniquely determines all the others.

3. Euclidean affine geometry.

The equations (2·7) can be interpreted as transformations of points if we regard x and y in (2·7) as the coordinates of two variable points in the same cartesian coordinate system. The set of all such point transformations for which the determinant $|p_i^j|$ is different from zero is uniquely determined by the fundamental differential form and is therefore an invariant. It is called *the affine group*. The set of all theorems stating properties which are unchanged by the transformations of the affine group is called the *affine Euclidean geometry*.

The affine group has a subgroup called the *displacement group* consisting of all point transformations represented by (2·7) in case the coordinates are orthogonal and the matrix $\| p_j^i \|$ is orthogonal with determinant $+1$. Any such transformation displaces any figure into a *congruent* figure. It leaves the fundamental differential form invariant.

The affine group also has a subgroup called the *similarity group* consisting of all point transformations of the form

$$y^i = qp_j^i \, x^j + p^i$$

in which the matrix $\| p_j^i \|$ is orthogonal. Any such transformation changes any figure into a similar figure and changes the fundamental form into one differing from it by a constant factor.

The theorems of Euclidean affine geometry constitute a significant subclass of the theorems of Euclidean geometry. This sub-class does not include theorems about the metric relations or such matters as similar triangles. But it does deal with the *straight lines*, i.e., the curves which have linear parameter representation

(3·1) $$y^i = a^i t + b^i$$

in cartesian coordinates. The equations (3·1) are solutions of the differential equations

(3·2) $$\frac{d^2 y^i}{dt^2} = 0.$$

Both (3·1) and (3·2) are unchanged in form by the linear transformations (2·7). But if we transform to arbitrary coordinates, we find

(3·3) $$\frac{d^2 y^i}{dt^2} = \frac{\partial^2 y^i}{\partial x^j \partial x^k} \frac{dx^j}{dt} \frac{dx^k}{dt} + \frac{\partial y^i}{\partial x^j} \frac{d^2 x^j}{dt^2},$$

and hence by (2·4) that (3·2) reduces to

(3·4) $$\frac{d^2 x^i}{dt^2} + \Gamma_{jk}^i \frac{dx^j}{dt} \frac{dx^k}{dt} = 0.$$

Hence the differential equations of the straight lines in arbitrary coordinates involve the components of affine connection in the form (3·4). As a corollary, it follows that if any transformation of coordinates is effected on the equations (3·4), they transform into a new set of equations of the same form in terms of the components of affine connection in the new coordinates.

The Euclidean affine geometry can be regarded as the theory of a fundamental affine connection whose components vanish in a particular coordinate system. The term "affine connection" in general is due to the fact that its theory is a generalization of the Euclidean affine geometry.

4. Euclidean vector analysis.

The orthogonal group is most compactly characterized by the condition (2·9). If we compare the laws of transformation for a covariant and a contravariant vector,

$$(4 \cdot 1) \qquad \overline{V}_i = V_a \frac{\partial x^a}{\partial \overline{x}^i}, \ \ \overline{V}^i = V^a \frac{\partial \overline{x}^i}{\partial x^a},$$

we see that the condition (2·9) is equivalent to the statement that *an orthogonal transformation has the same effect on the components of a covariant vector as on those of a contravariant vector.*

This explains why books on elementary vector analysis do not usually draw the distinction between covariant and contravariant vectors. For the operations with vectors which are usually considered in such books can all be interpreted by reference to components of the vectors in rectangular cartesian coordinate systems. This remark applies even if coordinates are dispensed with entirely, for in such cases components are taken by means of orthogonal projection or other processes depending on a Euclidean metric.

Since the determinant of a direct orthogonal transformation is unity, such a transformation has the same effect on the components of a relative vector of any weight as on the components of an absolute vector. Hence elementary vector analysis takes no account of the distinction between relative and absolute vectors. When it has to reckon with the distinction between right-handed and left-handed coordinate systems, it does so by special methods.

5. The customary geometrical interpretation of a vector as a "step" from an initial point x to a terminal point y may be justified by the following theorem: A point x, a contravariant vector A^i and a fundamental quadratic differential form determine a second point y uniquely which is such that *in any cartesian coordinate system*

$$(5 \cdot 1) \qquad y^i = x^i + A^i.$$

By way of proof we need only observe that if we transform to any other cartesian coordinate system $\bar{x}$, we find

$$\bar{x}^i = p_a^i x^a + p^i, \quad \bar{y}^i = p_a^i y^a + p^i, \quad \bar{A}^i = p_a^i A^a,$$

and therefore the formula (5·1) holds in the coordinate system $\bar{x}$ as well as in the coordinate system x. Since the totality of cartesian coordinate systems is uniquely determined by the fundamental quadratic form, the point y is uniquely determined by the point x, the vector A and the fundamental form. A slightly different formulation of the same result is that the differences of the coordinates of any two points transform like the components of a contravariant vector under transformations between any two cartesian systems of coordinates.

If U^i and V^i are contravariant vectors, $W^i = U^i + V^i$, and x is an arbitrary point given in cartesian coordinates, the points x, $x + U$, $x + W$, $x + V$ are the vertices of a parallelogram. Hence the usual parallelogram law for the addition of vectors follows. The parallelogram law is a Euclidean property of the vectors, because the system of all cartesian coordinate systems is an invariant of the Euclidean geometry.

6. Associated vectors and tensors.

For any contravariant vector V^i the fundamental quadratic form determines an associated covariant vector V_i by the formula

(6·1) $$V_i = g_{ij} V^j,$$

and for any covariant vector V_i it determines an associated contravariant vector V^i by the formula

(6·2) $$V^i = g^{ij} V_j.$$

If V_i is defined by (6·1) in terms of V^i, then V^i is the associated vector of V_i for

$$V^i = g^{ij} V_j = g^{ij} g_{aj} V^a = \delta_a^i V^a.$$

In Euclidean geometry in orthogonal coordinates, two associated vectors have the same components, and conversely, a covariant and a contravariant vector of the same weight which have the same components in an orthogonal coordinate system are associated.

If T_{ij} is an arbitrary tensor we can form two associated mixed tensors, by the formulas

(6·3) $$T^i_{\cdot j} = g^{ia} T_{aj}, \quad T_{i\cdot}^{\cdot j} = g^{ja} T_{ia},$$

and one associated contravariant tensor by the formulas

(6·4) $T^{ij} = g^{ai} g^{bj} T_{ab}.$

The dots are introduced in (6·3) because the relative order of the superscripts and subscripts i, j plays a rôle. The formulas

$$T_{ij} = g_{ia} g_{jb} T^{ab} = g_{ia} T^{a \cdot}_{\cdot j} = g_{ja} T^{\cdot a}_{i \cdot}.$$

and others of like character are now easily verified.

It is evident that this notion of tensors associated by means of a fundamental quadratic form can be extended to tensors of any order. It is used extensively in some of the books on Relativity. Some authors regard two associated vectors as a single object and thus speak of the covariant and contravariant components of the same vector, V. This usage is especially satisfactory in Euclidean geometry because we can associate the same geometric figure, the step, with either sort of vector.

In some problems it is desirable to represent the vector V (or the physical entity which the vector stands for) also by the covariant vector density $v_i = \sqrt{g} V_i$; or by the contravariant vector density, $v^i = \sqrt{g} g^{ia} V_a$. If we are dealing with the three dimensional case, we can also represent V by the contravariant tensor density $v^{ij} = \epsilon^{ijk} V_k$; or by the absolute contravariant tensor, $V^{ij} = \dfrac{\epsilon^{ijk}}{\sqrt{g}} V_k$; or by the absolute covariant tensor $V_{ij} = \sqrt{g} \epsilon_{ijk} V^k = \sqrt{g} \epsilon_{ijk} g^{kc} V_c$; and so on. Let us note that

$$g^{ia} g^{jb} V_{ij} = \sqrt{g} \epsilon_{ijk} g^{ia} g^{jb} g^{kc} V_c = \frac{\epsilon^{abc}}{\sqrt{g}} V_c = V^{ab}.$$

If we are dealing with orthogonal coordinates, all of the first four representations have the same components, namely (V_1, V_2, V_3). The other three representations all have the components

$$
\begin{array}{ccc}
0 & V_3 & -V_2 \\
-V_3 & 0 & V_1 \\
V_2 & -V_1 & 0
\end{array}
$$

7. Distance and scalar product.

The Euclidean distance from the initial point of a vector to its terminal point is

(7·1) $V = \sqrt{(V^1)^2 + (V^2)^2 + \ldots + (V^n)^2}$

in terms of the components of the vector in an orthogonal coordinate system. This quantity is called the *length* of the vector. It is a scalar which is a simultaneous invariant of the vector and the fundamental quadratic differential form of the Euclidean geometry. The expressions for its value in any other coordinate system are

$$(7\text{·}2) \qquad V = \sqrt{g_{ij}V^iV^j} = \sqrt{g^{ij}V_iV_j}.$$

For both expressions under the radical are scalars by § 11, Chap. II, and reduce to (7·1) for orthogonal coordinates. So, for example, in polar coordinates the length of a contravariant vector is

$$(7\text{·}3) \qquad \sqrt{(V^r)^2 + r^2\,(V^\theta)^2 + r^2\cos^2\theta\,(V^\phi)^2},$$

and the length of a covariant vector is

$$\sqrt{(V_r)^2 + \frac{1}{r^2}(V_\theta)^2 + \frac{1}{r^2\cos^2\theta}(V_\phi)^2}.$$

The scalar product which was defined in § 15, Chap. II, can be written in the following four forms

$$(7\text{·}4) \qquad A_iB^i = g_{ij}A^iB^j = g^{ij}A_iB_j = A^iB_i$$

in terms of associated vectors. Thus we are entitled to speak of the scalar product of two covariant vectors or of two contravariant vectors, meaning in each case a simultaneous invariant of two vectors and of the fundamental differential form.

It can be shown by familiar methods of analytic geometry, that an orthogonal coordinate system y can be found in which the y^1-axis has the same direction as A^i and the y^2-axis is such that B^i is in the same flat pencil of directions with the y^1 axis and the y^2 axis, i.e., such that the components of A^i and B^i reduce to $(\bar{A}, 0, 0, ..., 0)$ and $(\bar{B}^1, \bar{B}^2, 0, 0, ..., 0)$ respectively. The scalar product then becomes $\bar{A}\bar{B}^1$. In other words, it is the product of the lengths of the two vectors by the cosine of the angle between them. Since A_iB^i is a scalar this geometric interpretation is independent of the choice of coordinates. Hence we have

$$(7\text{·}5) \qquad \cos\theta = \frac{g_{ij}A^iB^j}{\sqrt{g_{ij}A^iA^j}\sqrt{g_{ij}B^iB^j}} = \frac{g^{ij}A_iB_j}{\sqrt{g^{ij}A_iA_j}\sqrt{g^{ij}B_iB_j}}$$
$$= \frac{A_iB^i}{\sqrt{A_iA^iB_jB^j}}$$

as the formula for the cosine of the angle between two vectors.

8. Area.

Another simultaneous invariant of two vectors and the fundamental form is the absolute scalar

$$(8\cdot1)\qquad C = g_{ab;ij}A^aB^bA^iB^j = \delta^{pq}_{ab}g_{ip}g_{jq}A^aB^bA^iB^j.$$

Here we have made use of (10·2), Chap. I. This scalar is connected with the outer product of the two vectors (§ 15, Chap. II) by the formula

$$(8\cdot2)\qquad C = \tfrac{1}{2}g_{ip}g_{jq}C^{ij}C^{pq}.$$

In case $n = 3$, C is the square of the length of an absolute vector which is connected with the relative vector c_i of § 15, Chap. II, by the equations

$$(8\cdot3)\qquad C_i = \sqrt{g}\,c_i = \tfrac{1}{2}\sqrt{g}\,\epsilon_{ijk}C^{jk} = \sqrt{g}\,\epsilon_{ijk}A^jB^k,$$

and is called the *vector product* of A and B. The components of the vector product are

$$C_1 = \sqrt{g}\,(A^2B^3 - A^3B^2),\quad C_2 = \sqrt{g}\,(A^3B^1 - A^1B^3),$$
$$C_3 = \sqrt{g}\,(A^1B^2 - A^2B^1).$$

To find the geometric meaning of these invariants we choose the same orthogonal coordinates as were used for a similar purpose in the last section. Then all the components of C^{ij} are zero except $C^{12} = -C^{21}$ which is equal in absolute value to the area of the parallelogram of the two vectors at the origin. The scalar C is the square of this area, and the length of the vector C_i is equal to this area. The components of C_i are $(0, 0, C^{12})$. Hence C_i has the direction of the x^3-axis.

Since C is a scalar we are able to infer that it is the square of the area of the parallelogram determined by the two vectors, irrespective of coordinate systems. Since perpendicularity is an invariant relation, the vector C_i is always orthogonal to the vectors A^i and B^i. We can also infer that the components of the tensor C^{ij} in any *cartesian coordinate system* are the areas of the projections of the parallelogram of A^i and B^i on the coordinate planes y^iy^j.

From the geometric interpretation of the outer and vector products we obtain the following formulas for the sine of the angle between two vectors

$$(8\cdot4)\quad \sin\theta = \frac{\sqrt{g_{ab;ij}A^aB^bA^iB^j}}{\sqrt{g_{ij}A^iA^j}\sqrt{g_{ij}B^iB^j}} = \frac{\sqrt{g^{ij}C_iC_j}}{\sqrt{g_{ij}A^iA^j}\sqrt{g_{ij}B^iB^j}}.$$

9. The tensor $g_{ab;ij}$ is used also in the formula for the area of a surface. To define a surface in a space of n dimensions (in which the coordinates may be either real or imaginary) we first define what is called a 2-cell by means of the parametric equations

$$(9 \cdot 1) \qquad x^i = \phi^i (t^1, t^2), \ 0 \leqslant t^1 \leqslant 1 \text{ and } 0 \leqslant t^2 \leqslant 1$$

with suitable conventions as to double points.

The area of a 2-cell is defined by the formula

$$(9 \cdot 2) \qquad A = \int_0^1 \int_0^1 \sqrt{g_{ab;ij} \frac{\partial x^a}{\partial t^1} \frac{\partial x^b}{\partial t^2} \frac{\partial x^i}{\partial t^1} \frac{\partial x^j}{\partial t^2}} \, dt^1 dt^2.$$

This formula is justified by the facts (1) that A is unaltered by transformations of the coordinates x, and (2) that A is unaltered by transformation of the parameters such that the 2-cell continues to be defined by equations analogous to (9·1). We shall not stop to prove these statements; they depend on the fact that the integrand is an invariant which is homogeneous of degree zero in dt^1 and dt^2.

The definition of 2-cell and surface and the discussion of the formulas for length and area generalize without difficulty to any number of dimensions. Thus for a k-cell, K, we have the k-dimensional volume

$$\int \dots k \dots \int \sqrt{g_{ab \dots d; ij \dots l} \frac{\partial x^a}{\partial t^1} \dots \frac{\partial x^c}{\partial t^k} \frac{\partial x^i}{\partial t^1} \dots \frac{\partial x^l}{\partial t^k}} \, dt^1 dt^2 \dots dt^k.$$

If $k = n$ this reduces to

$$\int \dots n \dots \int \sqrt{g} \, dx^1 dx^2 \dots dx^n.$$

10. First order differential parameters.

By specializing from a general vector to the gradient of a scalar, we can obtain at once a list of simultaneous invariants of a quadratic differential form and of one or more scalars. These simultaneous invariants are called differential parameters because of the way in which they were first used in differential geometry and mathematical physics.

Since the gradient of a scalar is a vector, the fundamental quadratic form determines for any scalar V, another scalar, the square of the magnitude of the gradient of V,

$$(10 \cdot 1) \qquad \Delta_1 (V) = g^{ij} \frac{\partial V}{\partial x^i} \frac{\partial V}{\partial x^j}.$$

This invariant is called the Lamé differential parameter of the first order. In orthogonal coordinates it reduces to

$$\left(\frac{\partial V}{\partial x^1}\right)^2 + \left(\frac{\partial V}{\partial x^2}\right)^2 + \left(\frac{\partial V}{\partial x^3}\right)^2,$$

and in polar coordinates to

$$\left(\frac{\partial V}{\partial r}\right)^2 + \frac{1}{r^2}\left(\frac{\partial V}{\partial \theta}\right)^2 + \frac{1}{r^2 \cos^2 \theta}\left(\frac{\partial V}{\partial \phi}\right)^2.$$

For the two scalars V and W, the fundamental quadratic form determines a scalar,

$$(10 \cdot 2) \qquad \Delta_1(V, W) = g^{ij}\frac{\partial V}{\partial x^i}\frac{\partial W}{\partial x^j},$$

the scalar product of their gradients. In orthogonal coordinates this is

$$\frac{\partial V}{\partial x^1}\frac{\partial W}{\partial x^1} + \frac{\partial V}{\partial x^2}\frac{\partial W}{\partial x^2} + \frac{\partial V}{\partial x^3}\frac{\partial W}{\partial x^3},$$

and in polar coordinates

$$\frac{\partial V}{\partial r}\frac{\partial W}{\partial r} + \frac{1}{r^2}\frac{\partial V}{\partial \theta}\frac{\partial W}{\partial \theta} + \frac{1}{r^2 \cos^2 \theta}\frac{\partial V}{\partial \phi}\frac{\partial W}{\partial \phi}.$$

Forming the vector product of the two gradients we obtain a contravariant vector,

$$(10 \cdot 3) \qquad \Theta^i(V, W) = \frac{1}{\sqrt{g}}\,\epsilon^{ijk}\frac{\partial V}{\partial x^j}\frac{\partial W}{\partial x^k},$$

and the square of the magnitude of this is

$$(10 \cdot 4) \qquad \frac{g_{ab}}{g}\,\epsilon^{ajk}\,\epsilon^{bpq}\frac{\partial V}{\partial x^j}\frac{\partial V}{\partial x^p}\frac{\partial W}{\partial x^k}\frac{\partial W}{\partial x^q}.$$

In three-dimensional orthogonal coordinates we have

$$\Theta^1(V, W) = \left(\frac{\partial V}{\partial x^2}\frac{\partial W}{\partial x^3} - \frac{\partial V}{\partial x^3}\frac{\partial W}{\partial x^2}\right),$$

and two other equations obtained by cyclic permutation of the indices.

11. Euclidean covariant differentiation.

The geometrical meaning of Euclidean covariant differentiation is self-evident. For since the components of affine connection all vanish in cartesian coordinates, the covariant derivative of any tensor T is *that tensor whose components in any cartesian coordinate system are the partial derivatives of the components of T in this* *coordinate system.* This is an instance of a very important general process in invariant theory. In case a basic invariant or set of invariants (e.g. the Euclidean fundamental form) determines an invariant class of coordinate systems, a properly chosen operation in one of these coordinate systems will give rise to new invariants.

12. The divergence.

The process of covariant differentiation can be used to form a number of important invariants which have no relation to the basic affine connection and therefore no relation to the Euclidean geometry. In such cases covariant differentiation is used merely as a systematic computing device for the elimination of second derivatives. For example, if v^i is a contravariant vector density, we have

$$v^i_{,j} = \frac{\partial v^i}{\partial x^j} + v^a\,\Gamma^i_{aj} - v^i\,\Gamma^a_{aj}.$$

If we set $i = j$ and sum, we find

$$(12 \cdot 1) \qquad\qquad v^i_{,i} = \frac{\partial v^i}{\partial x^i}.$$

Since the expression on the left of this equation is a scalar density, $\partial v^i/\partial x^i$ is a scalar density. It is called the *divergence* of the vector density v^i. It is obviously an invariant of the contravariant vector density alone and has nothing to do either with the affine connection or the Euclidean fundamental form.

The name divergence is also applied to a simultaneous invariant of a contravariant vector and the fundamental quadratic form. If V^i is an absolute contravariant vector, $\sqrt{g}\,V^i$ is a contravariant vector density. Hence,

$$(12 \cdot 2) \qquad\qquad \frac{\partial\,(\sqrt{g}\,V^i)}{\partial x^i}$$

is a scalar density and

$$\frac{1}{\sqrt{g}}\,\frac{\partial\,(\sqrt{g}\,V^i)}{\partial x^i}$$

is an absolute scalar. This is what we may call the *absolute divergence.* In orthogonal coordinates it reduces to

$$\frac{\partial V^1}{\partial x^1} + \frac{\partial V^2}{\partial x^2} + \dots + \frac{\partial V^n}{\partial x^n},$$

and is indistinguishable from the density divergence (12·1). But in

3-dimensional polar coordinates it is

$$\frac{1}{r^2 \cos \theta} \left(\frac{\partial}{\partial r} (r^2 V^r \cos \theta) + \frac{\partial}{\partial \theta} (r^2 V^\theta \cos \theta) + \frac{\partial}{\partial \phi} (r^2 V^\phi \cos \theta) \right),$$

whereas the density divergence is

$$\frac{\partial V^r}{\partial r} + \frac{\partial V^\theta}{\partial \theta} + \frac{\partial V^\phi}{\partial \phi}.$$

13. The Laplacian or Lamé differential parameter of the second order.

If V is an absolute scalar, its gradient $\partial V / \partial x^i$ is a covariant vector and hence

(13·1) $$\sqrt{g} \, g^{ia} \frac{\partial V}{\partial x^a}$$

is a contravariant vector density. Hence by the last section

(13·2) $$\Delta_2 (V) = \frac{1}{\sqrt{g}} \frac{\partial}{\partial x^i} \left(\sqrt{g} \, g^{ia} \frac{\partial V}{\partial x^a} \right) = g^{ij} V_{,\, i,\, j}$$

is an absolute scalar called the Lamé differential parameter of the second order. In orthogonal coordinates it reduces to

$$\Delta_2 (V) = \frac{\partial^2 V}{(\partial x^1)^2} + \frac{\partial^2 V}{(\partial x^2)^2} + \cdots + \frac{\partial^2 V}{(\partial x^n)^2},$$

which is known by various names such as the Laplacian of V, or the nabla of V, etc. In polar coordinates it becomes

$$\Delta_2 (\overline{V}) = \frac{\partial^2 \overline{V}}{\partial r^2} + \frac{1}{r^2} \frac{\partial^2 \overline{V}}{\partial \theta^2} + \frac{1}{r^2 \cos^2 \theta} \frac{\partial^2 \overline{V}}{\partial \phi^2} + \frac{2}{r} \frac{\partial \overline{V}}{\partial r} - \frac{\tan \theta}{r^2} \frac{\partial \overline{V}}{\partial \theta}$$

$$= \frac{1}{r^2} \left[\frac{\partial}{\partial r} \left(r^2 \frac{\partial \overline{V}}{\partial r} \right) + \frac{1}{\cos \theta} \frac{\partial}{\partial \theta} \left(\cos \theta \frac{\partial \overline{V}}{\partial \theta} \right) + \frac{1}{\cos^2 \theta} \frac{\partial^2 \overline{V}}{\partial \phi^2} \right].$$

For example, Poisson's differential equation is

$$\frac{\partial^2 V}{\partial x^2} + \frac{\partial^2 V}{\partial y^2} + \frac{\partial^2 V}{\partial z^2} = - 4\pi \rho,$$

in which ρ is a density. For transformation to arbitrary coordinates it is convenient to write this equation

$$\Delta_2 (V) = - 4\pi \frac{\rho}{\sqrt{g}} = - 4\pi \delta;$$

and so, in polar coordinates, Poisson's equation becomes

$$\frac{\partial^2 \overline{V}}{\partial r^2} + \frac{2}{r}\frac{\partial \overline{V}}{\partial r} + \frac{1}{r^2}\frac{\partial^2 \overline{V}}{\partial \theta^2} - \frac{\tan\theta}{r^2}\frac{\partial \overline{V}}{\partial \theta} + \frac{1}{r^2\cos^2\theta}\frac{\partial^2 \overline{V}}{\partial \phi^2} = -4\pi\delta.$$

14. The curl of a vector.

If V_i is an absolute covariant vector, its covariant derivative with respect to any affine connection is

$$V_{i,j} = \frac{\partial V_i}{\partial x^j} - V_a \Gamma_{ij}^a.$$

Since $V_{i,j} - V_{j,i}$ is a tensor it follows that

$$(14 \cdot 1) \qquad U_{ij} = \frac{\partial V_i}{\partial x^j} - \frac{\partial V_j}{\partial x^i} = \delta_{ij}^{ab}\frac{\partial V_a}{\partial x^b}$$

is a tensor which is an invariant of the given vector alone. We shall call this the Stokes tensor because of the rôle it plays in Stokes's theorem.

In case $n = 3$, we can multiply U_{ij} by the contravariant tensor density ϵ^{ijk} and on summing obtain the *contravariant vector density*,

$$(14 \cdot 2) \qquad u^i = \frac{1}{2}\epsilon^{ijk}U_{jk} = \epsilon^{ijk}\frac{\partial V_j}{\partial x^k},$$

which has the components

$$u^1 = \frac{\partial V_2}{\partial x^3} - \frac{\partial V_3}{\partial x^2}, \quad u^2 = \frac{\partial V_3}{\partial x^1} - \frac{\partial V_1}{\partial x^3}, \quad u^3 = \frac{\partial V_1}{\partial x^2} - \frac{\partial V_2}{\partial x^1}.$$

The absolute vector,

$$(14 \cdot 3) \qquad U^i = -\frac{1}{\sqrt{g}}u^i$$

is called the *curl* or the *rotation* or the *rotor* of the vector V_i.

15. Generalized divergence and curl.

If $a^{ij\ldots kl}$ is an alternating tensor density of order p, then

$$\frac{\partial a^{ij\ldots kl}}{\partial x^l}$$

is an alternating tensor density of order $p - 1$, which we call the *divergence* of a. The proof of this theorem is a direct generalization of § 12. For, by the formula for covariant differentiation

$$a_{,l}^{ij...kl} = \frac{\partial a^{ij...kl}}{\partial x^l} + a^{aj...kl} \Gamma_{al}^i + ... + a^{ij...al} \Gamma_{al}^k$$
$$+ a^{ij...ka} \Gamma_{al}^l - a^{ij...kl} \Gamma_{al}^a.$$

The last two terms cancel each other, and the other terms with Γ as a factor vanish because of the alternating character of $a^{ij...kl}$. Hence we have

$$\frac{\partial a^{ij...kl}}{\partial x^l} = a_{,l}^{ij...kl}.$$

The right-hand member is a tensor density of order $p - 1$ because it is obtained by contraction from a tensor density of order $p + 1$. Hence the left-hand member is a tensor density of order $p - 1$. It is alternating because of its form.

As a corollary we see, just as in § 12, that if $A^{ij...kl}$ is an alternating absolute tensor of order p

$$\frac{1}{\sqrt{g}} \frac{\partial (\sqrt{g} A^{ij...kl})}{\partial x^l}$$

is an absolute tensor of order $p - 1$. This may be called the *absolute divergence* of A.

16. If $T_{ij...k}$ is an absolute tensor of order p, then

(16·1) $$\frac{1}{p!} \delta_{ab...cd}^{ij...kl} \frac{\partial T_{ij...k}}{\partial x^l} = P_{ab...cd}$$

is an absolute alternating tensor of order $p + 1$. It is a generalization of the Stokes tensor which may be called the Stokes-Poincaré tensor because of the rôle it plays in the theorem of multiple integration obtained by Poincaré as a generalization of the theorem of Stokes. To prove its tensor character we consider the equation

$$\delta_{ab...cd}^{ij...kl} T_{ij...k,l} = \delta_{ab...cd}^{ij...kl} \left(\frac{\partial T_{ij...k}}{\partial x^l} - T_{pj...k} \Gamma_{il}^p - ... - T_{ij...p} \Gamma_{kl}^p \right).$$

Since $\Gamma_{jk}^i = \Gamma_{kj}^i$, all the terms in Γ cancel out when multiplied by the Kronecker delta and summed. Since the left-hand member of the equation represents the components of a tensor, the theorem follows.

In case T is itself an alternating tensor, the formula for the Stokes-Poincaré tensor becomes particularly interesting. For example, in case of an alternating tensor of the second order, A_{ij} we find

(16·2) $$\frac{\partial A_{ij}}{\partial x^k} + \frac{\partial A_{jk}}{\partial x^i} + \frac{\partial A_{ki}}{\partial x^j} = P_{ijk},$$

and in case of an alternating tensor of the third order A_{ijk}, we find the tensor

(16·3) $$\frac{\partial A_{ijk}}{\partial x^l} - \frac{\partial A_{jkl}}{\partial x^i} + \frac{\partial A_{kli}}{\partial x^j} - \frac{\partial A_{lij}}{\partial x^k} = P_{ijkl}.$$

It is obvious that

(16·4) $$p^i = \frac{1}{(n-1)!}\, \epsilon^{ijk\ldots m}\, P_{jk\ldots m}$$

is a vector density in case the number of coordinates n is one more than the number of subscripts of $P_{jk\ldots m}$. This generalizes (14·2) and we can obtain a generalization of the curl by introducing a factor $1/\sqrt{g}$ as in (14·3).

By inspection of (16·1) we see that

(16·5) $$\delta^{ij\ldots lm}_{ab\ldots cd}\frac{\partial P_{ij\ldots l}}{\partial x^m} = 0.$$

If we multiply (16·5) by $\epsilon^{ab\ldots d}$ and sum we find

(16·6) $$\frac{\partial p^i}{\partial x^i} = 0,$$

from which we infer that the divergence of a curl or a generalized curl is zero.

17. Historical remarks.

The first systematic treatment of Euclidean geometry by means of general coordinates seems to have been that of G. Lamé, *Leçons sur les fonctions inverses des transcendantes et les surfaces isothermes*, Paris, 1857, and *Leçons sur les coordonnées curvilignes*, Paris, 1859. He introduced the term differential parameter, which was adopted by E. Beltrami who generalized the differential parameters to arbitrary quadratic forms. A good account of the early work on differential parameters by Lamé, Jacobi, Chelini, Brioschi, and Codazzi is to be found in the memoir of Beltrami, *Memorie dell' Acc. delle Scienze dell' Istituto di Bologna*, Ser. II, vol. 8 (1868), p. 551 (*Collected works*, vol. 2, p. 74, Milan, 1904).

CHAPTER V

THE EQUIVALENCE PROBLEM

1. Riemannian geometry.

The theory of an arbitrary quadratic differential form

(1·1) $\qquad\qquad g_{ij}\,dx^i\,dx^j$, for which $g \neq 0$,

is often referred to as a *Riemannian geometry*. The object obtained by associating the space whose coordinates are x with a differential form (1·1) is called a *Riemannian space*. Thus the same space in the sense of § 1, Chap. II, can be the "bearer" of any number of Riemannian spaces. The quadratic differential form is also called a *Riemannian metric* since it leads to a definite system of measurement in the Riemannian space.

The tensors g_{ij} and g^{ij} are referred to as the fundamental covariant and the fundamental contravariant tensors respectively of the Riemannian geometry, and (1·1) is referred to as the fundamental differential form of the geometry.

A Euclidean space is a special Riemannian space, and a large class of formulas and theorems of general Riemannian geometry can be obtained by direct generalization from the corresponding Euclidean formulas when the latter are written, as we wrote them in the last chapter, in arbitrary coordinates. Thus the formula (7·2) of Chap. IV gives us the length of a vector, and formula (1·2) gives the length of any curve. The proof that these quantities are invariants is exactly the same for the general case as it is for the Euclidean special case. To call them Riemannian lengths is merely a matter of terminology.

In like manner we adopt the definitions of scalar and vector product in §§ 7 and 8, Chap. IV, for Riemannian geometry and let (7·5) define the Riemannian angle, between two vectors. It then follows that (8·4) is also a correct formula for Riemannian angle. Riemannian area and volume are defined by means of the formulas in § 9, Chap. IV. All of this amounts to finding geometrical interpretations for simultaneous invariants of a quadratic differential form and one or more vectors.

The formulas for covariant differentiation in arbitrary coordinates are the same for Riemannian geometry as for the Euclidean special case. So also are the formulas derived from them, such as those for the absolute divergence, the generalized divergence, the curl, and all the differential parameters. The formulas for divergence as a density and for the Stokes-Poincaré tensor are, of course, also available because they are independent of any metric.

2. The theory of surfaces.

The differential geometry of surfaces in a Euclidean space is an application of two-dimensional Riemannian geometry. A contravariant vector dx is tangent to a surface

$$(2\cdot 1) \qquad x^i = \phi^i (u, v)$$

provided

$$(2\cdot 2) \qquad dx^i = \frac{\partial \phi^i}{\partial u} du + \frac{\partial \phi^i}{\partial v} dv.$$

The Euclidean formula for length of this vector gives

$$(2\cdot 3) \qquad g_{ij} dx^i dx^j = E du^2 + 2F du\, dv + G dv^2$$

in which g_{ij} is the fundamental tensor of the Euclidean space and

$$(2\cdot 4) \qquad E = g_{ij} \frac{\partial \phi^i}{\partial u} \frac{\partial \phi^j}{\partial u}, \quad F = g_{ij} \frac{\partial \phi^i}{\partial u} \frac{\partial \phi^j}{\partial v}, \quad G = g_{ij} \frac{\partial \phi^i}{\partial v} \frac{\partial \phi^j}{\partial v}.$$

The quadratic differential form $(2\cdot 3)$ in the two variables, u and v, is the fundamental form of a two-dimensional Riemannian geometry.

In two dimensions the non-zero components of the Riemann-Christoffel tensor differ among themselves at most in sign. Hence this tensor has only one essential component, and may be replaced by the relative scalar of weight 2,

$$\tfrac{1}{4} \epsilon^{ij} \epsilon^{ab} R_{ijab},$$

or the absolute scalar,

$$K = \frac{\epsilon^{ij} \epsilon^{ab} R_{ijab}}{4 (EG - F^2)},$$

which is known as the *total* or *Gaussian curvature*.

The problem of the applicability of two surfaces can be completely solved in terms of the total curvature and its covariant derivatives. This is a corollary of the equivalence theorem which we prove below. For a detailed discussion of applicability on this basis one may consult Ricci, *Lezioni sulla teoria delle superficie*, Padoa, 1898; and for an equivalent discussion in terms of differential parameters, Eisenhart, *Differential Geometry*, New York, 1909, Chap. IX.

The relations of the surface to the three-dimensional space in which it is situated are not completely determined by (2·3) and require the consideration also of a second fundamental differential form. For the details of this theory we must refer the reader to books on differential geometry.

3. Spaces immersed in a Euclidean space.

The theory of a manifold of n dimensions in a Euclidean space of a greater number of dimensions is an application of Riemannian geometry of n dimensions. Inversely, a count of the unknown functions in the partial differential equations analogous to (2·4) shows that any Riemannian space of n dimensions may be regarded as immersed in a Euclidean space of $n(n+1)/2$ dimensions. A rigorous proof of this theorem is not yet available, but doubtless will be soon.

The least dimensionality of a Euclidean space in which a given Riemannian space can be immersed is obviously an invariant of the Riemannian space. The difference between this number and the dimensionality of the Riemannian space is called the *class* of the Riemannian space. So, for example, a Euclidean space is of class zero and a sphere is a Riemannian space of dimensionality two and class one. A beginning of a theory of this invariant has been made by G. Ricci, but it has not been carried very far as yet. For an account of this subject as well as of the more general question of Riemannian spaces immersed in other Riemannian spaces the reader is referred to Eisenhart, *Riemannian Geometry*, Princeton, 1926, Chap. V.

The representation of a Riemannian space as a manifold in a Euclidean space underlies the symbolic method of H. Maschke for the theory of invariants of a quadratic differential form. For lack of space, we shall have to dismiss this subject with references to Maschke's first paper in the *Trans. Amer. Math. Soc.* vol. I (1900), p. 197 and to the recent paper by L. Ingold, *ibid.* vol. 27 (1925), p. 574.

4. Condition that a Riemannian space be Euclidean.

It is evident that the curvature tensor of a Riemannian space vanishes identically if the space is Euclidean. It is also true that a Riemannian space is Euclidean if the curvature tensor vanishes.

We shall prove this by showing that if the curvature tensor vanishes a coordinate system can be found in which the components of affine connection vanish identically. We have already proved that any affine connection whose components vanish identically in some coordinate system is the affine connection of a Euclidean space. If the components of affine connection are Γ^{i}_{jk} in an arbitrary coordinate system and are zero in a coordinate system y we must have

$$\frac{\partial^2 y^i}{\partial x^j \partial x^k} - \Gamma^{p}_{jk} \frac{\partial y^i}{\partial x^p} = 0,$$

according to the law of transformation (10·1), Chap. III. Otherwise said, a transformation from the coordinates x to the required coordinates y is determined by any set of n independent functions $y^1, y^2, \ldots, y^n$ which are solutions of

$$(4\cdot1) \qquad \frac{\partial^2 y}{\partial x^j \partial x^k} - \Gamma^{p}_{jk} \frac{\partial y}{\partial x^p} = 0.$$

The second order equations (4·1) are equivalent to the first order equations

$$(4\cdot2) \qquad \frac{\partial y}{\partial x^j} = u_j, \qquad \frac{\partial u_j}{\partial x^k} = \Gamma^{p}_{jk} u_p.$$

These first order equations are of the general form*

$$(4\cdot3) \qquad \frac{\partial Z^a}{\partial x^i} = \Psi^{a}_{i}(Z, x)$$

in which the index a runs from 1 to R and the index i from 1 to n. The equations (4·3) reduce to (4·2) if we set $Z^1 = y$, $Z^2 = u_1$, $\ldots$, $Z^R = u_n$.

The equations (4·3) can also be written as the total differential equations,

$$(4\cdot4) \qquad dZ^a = \Psi^{a}_{i} dx^i,$$

the problem being to find functions $Z(x)$ whose total differentials are given by (4·4). The equations (4·3) or (4·4) are said to be *completely integrable* if for any set of values $(x_0^1, x_0^2, \ldots, x_0^n)$ and $(Z_0^1, Z_0^2, \ldots, Z_0^R)$ for which the functions Ψ are analytic there exists one and only one set of R functions $Z(x)$ which satisfy (4·3) and take on the

---

* In this chapter we shall have to deal with expressions in which the indices do not always run over the same range. Any index which does not run from 1 to n will be denoted by a Greek letter.

initial values Z_0 for $x = x_0$. It was proved by J. C. Bouquet* that (4·3) are completely integrable if and only if the "integrability conditions"

$$(4·5) \qquad \frac{\partial \Psi_i^{\cdot \alpha}}{\partial x^j} + \frac{\partial \Psi_i^{\cdot \alpha}}{\partial Z^\beta} \Psi_j^{\cdot \beta} = \frac{\partial \Psi_j^{\cdot \alpha}}{\partial x^i} + \frac{\partial \Psi_j^{\cdot \alpha}}{\partial Z^\beta} \Psi_i^{\cdot \beta}$$

are satisfied identically in the Z's and x's. For the proof of this theorem we must refer the reader either to Bouquet's article or to text-books on differential equations.

In the special case in which (4·3) becomes (4·2) the integrability conditions (4·5) reduce to

$$(4·6) \qquad \Gamma_{ij}^p u_p = \Gamma_{ji}^p u_p,$$

and

$$(4·7) \qquad B_{jkl}^p u_p = 0.$$

The conditions (4·6) are satisfied identically because of the symmetry of the components of affine connection. When the curvature tensor vanishes identically it therefore follows that the equations (4·2) are completely integrable. There therefore exists a solution of (4·2) in which y and $u_1, u_2, ..., u_n$ are given arbitrary initial values.

If we require that the origin of the cartesian coordinates shall be the point x_0, then we have $y = 0$ for $x = x_0$ and it is possible to choose n linearly independent sets of initial values of $u_1, u_2, ..., u_n$. We thus get n independent solutions of (4·1),

$$(4·8) \qquad y^i = A^{(i)}(x),$$

which define a transformation from the arbitrary coordinates x to the required coordinates y in which the components of affine connection vanish.

According to the definitions in § 2, Chap. IV, the coordinates y are cartesian. Indeed, the differential equations (2·5) of that section are identical with those at the beginning of this section.

5. If we substitute $A(x)$, a component of a scalar, for y in the left-hand member of (4·1), the result is the covariant derivative of the gradient, $\partial A / \partial x^i$. Hence if A is a solution of (4·1) in the coordinate system x, and $\bar{A}$ is the component of the scalar A in the coordinate system $\bar{x}$, $\bar{A}$ is a solution of the equations into which (4·1) transform when x is transformed to $\bar{x}$.

---

* *Bull. Sci. Math. et Astron.* (1), vol. 3 (1872), p. 265.

A particular cartesian coordinate system is determined by the n sealars
$$A^{(1)}(x), \ A^{(2)}(x), \ \ldots, \ A^{(n)}(x)$$
according to the formulas (4·8). And the same cartesian coordinate system is determined by the transformation
$$y^i = \bar{A}^i(\bar{x})$$
in which each $\bar{A}^{(i)}(\bar{x})$ is the component in $\bar{x}$ of the scalar $A^{(i)}(x)$.

The fact that all cartesian coordinate systems are linearly related to any one corresponds to the fact, obvious from the discussion in § 4, that all solutions of (4·1) are given by the formula
$$(5·1) \qquad\qquad p_i A^{(i)}(x) + p$$
in which the p's are arbitrary constants. The set of scalars (5·1) is an invariant of the Euclidean affine connection. Each of these scalars when equated to zero represents a hyperplane.

6. The equivalence problem.

The problem which we have just solved for the Euclidean geometry can be regarded as a special case of the following more general problem. Suppose that two sets of basic invariants $I_1, I_2, \ldots$, and $J_1, J_2, \ldots$, are given in terms of their components in a given coordinate system, does there exist a point transformation* $\bar{x}^i = f^i(x)$ which carries the one set of invariants into the other? If there does, the two sets of invariants are said to be *equivalent*. We have proved that two quadratic differential forms are equivalent if their curvature tensors vanish.

For quadratic differential forms in general the question may be stated as follows: If
$$(6·1) \qquad\qquad g_{ij}dx^i dx^j \quad \text{and} \quad \bar{g}_{ij}d\bar{x}^i d\bar{x}^j$$
are the components of two quadratic forms *in the same coordinate system* does there exist a point transformation
$$(6·2) \qquad\qquad x^i = x^i(\bar{x})$$
which carries the one quadratic form into the other? This is the same as the question whether there exists a set of $n^2 + n$ functions of $\bar{x}$, $x^i(\bar{x})$ and $u_j^i(\bar{x})$, such that
$$(6·3) \qquad\qquad \frac{\partial x^i}{\partial \bar{x}^j} = u_j^i,$$
and
$$(6·4) \qquad\qquad \bar{g}_{ij} = g_{pq} u_i^p u_j^q.$$

* The distinction between a transformation of points and a transformation of coordinates was brought out in § 2, Chap. II.

These equations imply (9·3), Chap. III, which we may now write

$$(6·5) \qquad \frac{\partial u^i_j}{\partial \bar{x}^k} = \overline{\Gamma}^p_{jk} u^i_p - \Gamma^i_{pq} u^p_j u^q_k,$$

in which

$$(6·6) \qquad \Gamma^i_{jk} = \frac{1}{2} \cdot g^{ai} \left(\frac{\partial g_{aj}}{\partial x^i} + \frac{\partial g_{ia}}{\partial x^j} - \frac{\partial g_{ij}}{\partial x^a} \right).$$

Since (6·3), (6·4) and (6·5) together imply (6·3) and (6·4), we see that (6·3) and (6·4) have a solution of the form required if and only if (6·3), (6·4) and (6·5) have a solution of this form.

If we replace the variables x^i and u^i_j by $Z^1, Z^2, \ldots, Z^R$ (where $R = n^2 + n$) and replace $\bar{x}^i$ by $x^1, x^2, \ldots, x^n$, then the equations (6·3) and (6·5) reduce to the form (4·3) and the equations (6·4) are of the form $F_\lambda (Z, x) = 0$.

7. A lemma on mixed systems.

Thus in order to solve the equivalence problem we need to find the conditions under which there exist solutions of a mixed system of differential equations,

$$(7·0) \qquad \frac{\partial Z^a}{\partial x^i} = \Psi^a_i (Z, x), \quad (a = 1, 2, \ldots, R)$$

and functional relations,

$$(7·1) \qquad F^{(0)}_\lambda (Z, x) = 0. \quad (\lambda = 1, 2, \ldots, L)$$

If we differentiate (7·0) with respect to x, substitute in

$$\frac{\partial^2 Z^a}{\partial x^i \partial x^j} = \frac{\partial^2 Z^a}{\partial x^j \partial x^i},$$

and replace $\partial Z^a / \partial x^i$ by its value according to (7·0) we obtain the equations (4·5) which must be satisfied by all solutions of (7·0). The solutions of (7·0) must also satisfy all the equations obtained by differentiating (7·1) with respect to x and eliminating the derivatives $\partial Z^a / \partial x^i$ by means of (7·0). Let us represent the set of all equations obtained by these two processes by

$$(7·11) \qquad F^{(1)}_\mu (Z, x) = 0.$$

That is to say, (7·11) includes (4·5) and all the equations obtained by differentiation of (7·1) and elimination by means of (7·0). These equations may be modified in a variety of ways by combining them with (7·1). For the purpose of our argument (7·11) may equally well represent any set of equations which when combined with (7·1) have the same solutions as (7·1) and (7·11).

It is also evident that all solutions of (7·0) and (7·1) must satisfy the equations

(7·12) $F_\nu^{(2)}(Z, x) = 0,$

which are obtained by differentiating (7·11) with regard to x and eliminating $\partial Z^\alpha/\partial x^i$ by means of (7·0). Again, (7·12) may equally well represent any set of equations which when combined with (7·1) and (7·11) have the same solutions as (7·1), (7·11) and (7·12).

By repeating this process we get a sequence of relations

(7·13) $F_\rho^{(3)}(Z, x) = 0,$

(7·14) $F_\sigma^{(4)}(Z, x) = 0,$

$$\vdots$$

and so on, all of which are necessary conditions on the solutions of (7·0) and (7·1).

From the equations $F_\lambda^{(0)}(Z, x) = 0$ we can pick out a certain number which are independent and such that the rest are dependent on them. To this set we can add certain ones of the set $F_\mu^{(1)}(Z, x) = 0$ so that the equations of the resulting set are independent and all equations in the set $F_\mu^{(1)}(Z, x) = 0$ are dependent on them. If this process is continued we obtain, after a finite number N of steps, either an incompatible* set of equations or one such that all solutions of it satisfy all the remaining equations of the sequence. For it is impossible to have an infinite set of independent equations in a finite number of variables.

In the case of incompatibility there are no solutions of (7·0) and (7·1) because the necessary conditions $F_\sigma^{(\lambda)}(Z, x) = 0$ are not all satisfied. In the other case there is a first number N such that the equations

(7·2) $F_\lambda^{(0)}(Z, x) = 0,\quad F_\mu^{(1)}(Z, x) = 0,\ \ldots,\ F_\sigma^{(N)}(Z, x) = 0$

are compatible and any solution of them satisfies

(7·3) $F_\tau^{(N+1)}(Z, x) = 0$

also. By the process described in the last paragraph we found a set of independent equations which is equivalent to (7·2). These we denote by

(7·4) $G^\rho(Z, x) = 0.\quad (\rho = 1, 2, \ldots, M \leqslant R)$

---

* The case of incompatibility includes that in which an equation of the form $F(x) = 0$ is deducible from the set, for in this case there is no solution $Z(x)$ in which the x's are independent.

The condition that (7·4) are independent is equivalent* to the non-vanishing of one of the functional determinants of the G's as functions of M of the Z's. By an unessential change of notation the Z's can be so arranged that this determinant is

(7·5)
$$\frac{\partial (G^1, G^2 \dots G^M)}{\partial (Z^1, Z^2, \dots Z^M)} \neq 0.$$

On account of (7·5) we can solve (7·4) for the first M Z's as functions of the remaining Z's and the x's,

(7·6)
$$Z^a = \xi^a (Z^{M+1}, \dots, Z^R, x). \quad (a = 1, 2, \dots, M)$$

Within the region of convergence of this solution the equations (7·6) are completely equivalent to (7·2). Hence by the method of forming the sequence $F_\lambda^{(\sigma)}$ the equations (7·3) and (7·2) imply the relations

(7·7)
$$\Psi_i^a = \frac{\partial \xi^a}{\partial Z^\sigma} \Psi_i^\sigma + \frac{\partial \xi^a}{\partial x^i} \quad (a = 1, 2, \dots, M)$$

in which the index σ is summed from $M + 1$ to R.

When we substitute (7·6) in the equations (7·0) we obtain

(7·8)
$$\frac{\partial \xi^a}{\partial x^i} + \frac{\partial \xi^a}{\partial Z^\sigma} \frac{\partial Z^\sigma}{\partial x^i} = [\Psi_i^a], \quad (a = 1, 2, \dots, M)$$

and

(7·9)
$$\frac{\partial Z^\sigma}{\partial x^i} = [\Psi_i^\sigma], \quad (\sigma = M + 1, \dots, R)$$

in which the square brackets represent the function of x and Z^{M+1}, $\dots$, Z^R obtained by substituting (7·6) in the expression inside the bracket. The problem of solving (7·0) and (7·1) is now reduced to that of solving the differential equations (7·9) because any solutions of (7·9) satisfy (7·8) on account of (7·7).

The integrability conditions of (7·9) are

(7·10)
$$\frac{\partial [\Psi_i^\sigma]}{\partial x^j} + \frac{\partial [\Psi_i^\sigma]}{\partial Z^\rho} [\Psi_j^\rho] = \frac{\partial [\Psi_j^\sigma]}{\partial x^i} + \frac{\partial [\Psi_j^\sigma]}{\partial Z^\rho} [\Psi_i^\rho].$$
$$(\sigma, \rho, = M + 1, \dots, R)$$

The left-hand member of this is

$$\left[\frac{\partial \Psi_i^\sigma}{\partial x^j} + \frac{\partial \Psi_i^\sigma}{\partial Z^a} \frac{\partial \xi^a}{\partial x^j} \right] + \left[\frac{\partial \Psi_i^\sigma}{\partial Z^\rho} + \frac{\partial \Psi_i^\sigma}{\partial Z^a} \frac{\partial \xi^a}{\partial Z^\rho} \right] [\Psi_j^\rho],$$

---

* Cf. Osgood, *Lehrbuch der Funktionentheorie*, vol. 2, Chap. II, § 23, p. 122 (Leipzig, 1924).

which by (7·7) reduces to

$$\left[\frac{\partial \Psi_i^\sigma}{\partial x^j} + \frac{\partial \Psi_i^\sigma}{\partial Z^\alpha} \Psi_j^\alpha + \frac{\partial \Psi_i^\sigma}{\partial Z^\rho} \Psi_j^\rho \right],$$

which is the result of substituting (7·6) in the left-hand member of (4·5). Hence the equations (7·10) are what some of the equations $F_\mu^{(1)}(Z, x) = 0$ reduce to when (7·6) is substituted in. Hence they are satisfied identically, and the equations (7·9) are completely integrable. This proves the theorem:

A necessary and sufficient condition that the equations (7·0) and (7·1) may be satisfied by Z's which are functions of the independent variables x is that there shall exist a number N (a whole number or zero) such that the equations (7·2) are compatible and such that every solution Z (x) of them satisfies (7·3).

The solution of (7·0) and (7·1) is reduced by the substitution (7·6) to the solution of a completely integrable set of equations (7·9) except in the case in which $M = R$; in this case the equations (7·2) completely determine the functions Z.

8. Equivalence theorem for quadratic differential forms.

In order to apply the theorem of § 7 to the case of quadratic differential forms we first observe that when the differential equations (7·0) are replaced by (6·3) and (6·5) the integrability conditions (4·5) reduce to

$$(8·1) \qquad \bar{B}_{jkl}^i u_i^p = B_{qrs}^p u_j^q u_k^r u_l^s.$$

Since $g_{ij,k} = 0$, the equations obtained by differentiating (6·4) and eliminating derivatives by means of (6·3) and (6·5) are all satisfied identically. Hence the equations (8·1) take the place of (7·11). The result of differentiating with respect to $\bar{x}^m$ and eliminating by means of (6·3) and (6·5) is

$$(8·2) \qquad \bar{B}_{jkl,m}^i u_i^p = B_{qrs,t}^p u_j^q u_k^r u_l^s u_m^t$$

which corresponds to (7·12). If this process is repeated we obtain the law of transformation of the second covariant derivative, $B_{jkl,m,p}^i$ of the curvature tensor; and so on. Hence the theorem of § 7 gives the theorem:

A necessary and sufficient condition for the equivalence of two quadratic differential forms $g_{ij}dx^i dx^j$ and $\bar{g}_{ij}d\bar{x}^i d\bar{x}^j$ is the existence of a whole number N such that (1) the first N sets of equations in the *sequence of sets of equations* (6·4), (8·1), (8·2), ..., *which embody the laws of transformation of the quadratic form itself, of the curvature tensor, and of the successive covariant derivatives of the curvature tensor, shall be compatible equations for the variables* x^i *and* u^i_j *as functions of the independent variables* $\bar{x}$, *and* (2) *all solutions of these equations shall satisfy the* $(N + 1)$*st set of equations in the sequence.*

This theorem provides a definite process by means of which the equivalence or non-equivalence of two quadratic differential forms can be determined in a finite number of steps*. The most general case is that in which the equations (6·4) and (8·1) are either incompatible or else completely determine the functions x^i and u^i_j. If the two differential forms are to be equivalent, these functions must satisfy (8·2). Otherwise the two forms are not equivalent. Thus, in general, a quadratic differential form is characterized by its curvature tensor.

The equations $$\bar{R}_{ijkl} = R_{pqrs}\, u^p_i\, u^q_j\, u^r_k\, u^s_l$$

are consequences of (8·1) and (6·4) and, inversely, these equations and (6·4) imply (8·1). In a like manner (6·4) and the law of transformation of each successive covariant derivative of the covariant curvature tensor R (§ 18, Chap. III) are equivalent to (6·4) and the law of transformation of the corresponding repeated covariant derivative of the curvature tensor B. We may therefore state the equivalence theorem in terms of the multilinear forms defined in § 19, Chap. III, as follows:

A necessary and sufficient condition for the equivalence of two quadratic differential forms G_2 *and* $\bar{G}_2$ *is that there shall exist a number* N *such that the laws of transformation implied by the first* N *equations of the sequence,*

$$G_2 = \bar{G}_2,\ G_4 = \bar{G}_4,\ G_5 = \bar{G}_5,\ ...,$$

shall be consistent equations for the determination of the functions x^i *and* u^i_j *and that all solutions of these equations shall satisfy the law of transformation implied by the* $(N + 1)$*st equation of this sequence.*

9. Equivalence of affine connections.

We may note in passing that the question of the equivalence of any two affine connections Γ^i_{jk} and $\bar{\Gamma}^i_{jk}$ (whether related to quadratic

---

* One of these steps, however, is the process of recognizing whether or not a number of functional relations imply another given functional relation.

forms or not) is answered by the argument in §§ 7 and 8. The equivalence of symmetric affine connections depends upon (6·3) and (6·5) in which the Γ's and $\overline{\Gamma}$'s are given functions subject only to the conditions

$$\Gamma^i_{jk} = \Gamma^i_{kj} \text{ and } \overline{\Gamma}^i_{jk} = \overline{\Gamma}^i_{kj}.$$

Equations (6·3) and (6·5) are of the form (7·0), and there are no equations (7·1). The equations (8·1), (8·2), etc., arise just as in § 8. Hence we have that a necessary and sufficient condition for the equivalence of two affine connections is the existence of a number N such that (1) the equations (8·1), (8·2), ..., (8·N) which embody the laws of transformation of the curvature tensor and of its first $N - 1$ covariant derivatives shall be compatible equations for the determination of the functions x^i and u^i_j of $\bar{x}$ and that (2) all solutions of these equations shall satisfy the equations (8·$N + 1$) which embody the law of transformation of the Nth covariant derivative of the curvature tensor.

10. Automorphisms of a quadratic differential form.

In case the equations (6·4), (8·1), ... (8·N) have a unique solution and this satisfies (8·$N + 1$), it uniquely determines the transformation (6·2) of G_2 into $\overline{G}_2$. It is the same, of course, if the number of solutions is finite. But if a continuous variation of the solution is possible, then the actual transformations are found, according to § 7, by solving a completely integrable set of differential equations. They will therefore be functions of a number of parameters, the constants of integration.

Since the resultant of one transformation of G_2 into $\overline{G}_2$ followed by the inverse of another is a transformation of G_2 into itself, it follows that there must be a continuous or mixed group of transformations of G_2 into itself in the case which we are considering; and this "group of automorphisms" of G_2 must have a finite number of parameters.

In case the quadratic differential form is Euclidean, the group of automorphisms is the set of all orthogonal transformations and includes the displacement group (§ 3, Chap. IV).

11. Another simple case is that in which the covariant curvature tensor satisfies the condition

(11·1) $R_{ijkl} = K g_{ij;kl} = K(g_{\alpha}g_{jl} - g_{il}g_{jk}).$

The factor of proportionality must be an absolute scalar since $g_{ij;kl}$ is a covariant tensor of the fourth order (§ 10, Chap. I, and § 11, Chap. II).

The law of transformation of $g_{ij;kl}$ is a simple algebraic combination of that of g_{ij}. Hence if we have two quadratic forms G_2 and $\bar{G}_2$ for which (11·1) is satisfied, K being the same scalar in each case, the relation

$$\bar{R}_{ijkl} = R_{pqrs}\, u_i^p\, u_j^q\, u_k^r\, u_l^s$$

is satisfied whenever

$$\bar{g}_{ij} = g_{pq}\, u_i^p\, u_j^q$$

is satisfied. Hence the conditions of the theorem of § 8 are satisfied and the two quadratic differential forms are equivalent. It can be proved that in case $n > 2$, K is a constant (cf. Eisenhart, *Riemannian Geometry*, § 26). When K is constant the spaces satisfying (11·1) are said to be of *constant curvature*.

12. Equivalence theorem in terms of scalars.

We have seen in § 22, Chap. III, that in the case in which the laws of transformation of G_2, G_4, ..., G_N suffice for the unique determination of u_j^i as functions of the components of the tensors a number of absolute scalars, $S^{(\alpha)}$, can be found which are rational functions of g_{ij} and their partial derivatives. Whether the scalar invariants are obtained by this or by some other process, it is obviously a necessary condition for the equivalence of G_2 and $\bar{G}_2$ that each scalar invariant of G_2 when evaluated at a point x shall be equal to the corresponding scalar invariant of $\bar{G}_2$ when the latter is evaluated at the corresponding point $\bar{x}$. If this condition is satisfied and if there are n of these scalars whose components are independent functions of x, n of the equations

(12·1) $S(x) = \bar{S}(\bar{x})$

can be solved so as to determine a transformation

(12·2) $x^i = f^i(\bar{x})$

which carries the one set of n scalars into the other. That is to say, the equations (12·2) must satisfy all the equations (12·1).

Now let us suppose that N is the smallest number such that there are n independent scalars formed from the algebraic invariants of the sequence of algebraic forms, G_2, G_4, ..., G_N. If the corresponding scalars obtained from $\bar{G}_2$ are not equal to these, the two quadratic differential forms are not equivalent. If the corresponding scalars obtained from $\bar{G}_2$ are equal, we have (12·1) and (12·2), and also define a set of functions u^i_j by the equation

(12·3)
$$u^i_j = \frac{\partial f^i}{\partial \bar{x}^j}.$$

These functions may or may not appear as the unique solutions of the first $(N-2)$ sets of equations in the sequence,

(12·4)
$$\bar{g}_{ij} = g_{ab}\, u^a_i\, u^b_j$$
$$\bar{R}_{ijkl} = R_{abcd}\, u^a_i\, u^b_j\, u^c_k\, u^d_l$$
$$\vdots \qquad \vdots$$

But the n equations (12·1) and hence the equations (12·2) are consequences of the first $N-2$ sets of equations (12·4). The $(N-1)$st set of equations in (12·4) is found by differentiating the $(N-2)$nd set and eliminating the first derivatives by means of the equations (6·3) and (6·5). Hence the equations (12·3) must be deducible from the first $N-1$ sets of equations in (12·4). It follows that the first $N-1$ sets of equations in (12·4) include equations sufficient to determine u^i_j uniquely. Hence a complete set of absolute invariants of $G_2, G_4, \ldots, G_{N+1}$ exist and, moreover, are equal to the corresponding invariants of $\bar{G}_2, \bar{G}_4, \ldots, \bar{G}_{N+1}$, if and only if, the first $N-1$ sets of equations in (12·4) are compatible. Hence the following theorem holds: *If there is a number N such that there are n scalars formed from the absolute invariants of the sequence of algebraic forms $G_2, G_4, \ldots, G_N$ which are independent as functions of x, without this being the case if G_N be omitted from the sequence, then a necessary and sufficient condition for the equivalence of G_2 and $\bar{G}_2$ is that a complete set of absolute invariants of $G_2, G_4, \ldots, G_N, G_{N+1}$ shall be equal to the corresponding absolute invariants of $\bar{G}_2, \bar{G}_4, \ldots, \bar{G}_N, \bar{G}_{N+1}$.*

This theorem obviously has a more restricted domain of application than that of § 8 because it does not cover the cases in which the transformation is to be determined by solving a completely integrable set of differential equations. In this case it is clear that no matter how far the sequence of algebraic forms is continued we do not obtain n independent scalar invariants.

13. Historical remarks.

The equivalence problem was first formulated and solved by E. B. Christoffel, *Journal für die reine und angew. Math.* (Crelle), vol. 70 (1869), pp. 46 and 241 (also *Collected Works*, end of vol. 1).

His form of the theorem is essentially that given in § 12, above. The more general equivalence theorem of § 8 is to be found in § 17 of the *Cambridge Tract* No. 9 (1908) by J. E. Wright with the same title as this tract. The underlying theorem on differential equations in § 7 was formulated by O. Veblen and J. M. Thomas, *Annals of Math.* vol. 27 (1926), p. 288. The proof makes use of a known method of reducing a general set of equations (7·0) to a completely integrable set, but the exact formulation of the conditions of compatibility seems to be new. It can be used to prove a considerable number of analogous equivalence theorems.

CHAPTER VI

NORMAL COORDINATES

1. Affine geometry of paths.

The theory of an arbitrary affine connection, is called *affine differential geometry* or, more briefly, affine geometry. Thus, the theory of covariant differentiation in §§ 10 to 14, and the theory of the curvature tensor in §§ 16 and 17, Chap. III, belong to affine geometry. The object obtained by associating a space of any number of dimensions with an affine connection is called an *affine space*. In case the components of affine connection are Christoffel symbols, that is to say, in case the fundamental affine connection is such that there exists a symmetric covariant tensor g_{ij} satisfying the condition

$$(1\cdot1) \qquad\qquad g_{ij,\,k} = 0,$$

the affine geometry of paths reduces to a Riemannian geometry. The conditions which the components of affine connection must satisfy in order that this shall happen can be stated in terms of the curvature tensor and its covariant derivatives (cf. Eisenhart and Veblen, *Proc. Nat. Ac. Sc.*, vol. VIII (1922), p. 19).

In case there exists a coordinate system y in which the components of affine connection are all zero the affine space is said to be *flat*. Such a space may always be regarded as a Euclidean space, because in the coordinates y (1·1) reduces to

$$\frac{\partial g_{ij}}{\partial y^k} = 0$$

and hence the components of affine connection are the Christoffel symbols of any quadratic differential form with constant coefficients in this coordinate system. It is an obvious consequence of § 4, Chap. V, that an affine space is flat if and only if the curvature tensor vanishes identically.

The fundamental geometric figure of an affine space is the system of curves which satisfy the differential equations

$$(1\cdot2) \qquad\qquad \frac{d^2 x^i}{dt^2} + \Gamma^i_{jk}\,\frac{dx^j}{dt}\,\frac{dx^k}{dt} = 0,$$

in which the coefficients are the components of affine connection. In the case of a flat space these curves are the straight lines ((3·4),

Chap. IV). In the general case, as we shall prove below, they have the local* property that any two points may be joined by one and only one curve of the system. On account of this property we shall call them *paths*, for they serve as a means of finding one's way about the affine space.

In the Riemannian case the paths are the *geodesics* or extremal curves of the integral of length

$$\int \sqrt{g_{ij} \, dx^i \, dx^j}.$$

We shall not give the proof of this theorem because we do not use it in the sequel and because it is to be found in almost every book on Differential Geometry or Relativity. Instead, we shall develop a few of the general properties of the paths of an affine space, i.e. as much of the *affine geometry of paths* as is required for the discussion of the invariant coordinate systems which are the subject of this chapter.

2. If we make a transformation of coordinates,

$$\bar{x}^i = f^i (x),$$

the differential equations (1·2) become

(2·1) $$\frac{d^2 \bar{x}^i}{dt^2} + \bar{\Gamma}^i_{jk} \frac{d\bar{x}^j}{dt} \frac{d\bar{x}^k}{dt} = 0,$$

in which the coefficients Γ are the components of affine connection in the coordinate system $\bar{x}$. In other words, the differential equations (1·2) are invariant in form under transformations of coordinates. Moreover, if

(2·2) $$x^i = \phi^i (t)$$

is a solution of (1·2), then

$$\bar{x}^i = f^i (\phi^1 (t), \, \ldots, \, \phi^n (t))$$

is a solution of the transformed equations (2·1) and represents the same curve as (2·2). Hence the system of curves which satisfy (1·2) is uniquely determined by the affine connection and hence is an invariant of the affine connection.

---

* The adjective "local" implies that any point may be enclosed in an n-cell within which the property holds good.

To solve (1·2) formally we differentiate them successively and find the following sequence of equations which must be satisfied by any solution of (1·2):

(2·32)
$$\frac{d^3 x^i}{dt^3} + \Gamma^i_{abc} \frac{dx^a}{dt} \frac{dx^b}{dt} \frac{dx^c}{dt} = 0$$

(2·33)
$$\frac{d^4 x^i}{dt^3} + \Gamma^i_{abcd} \frac{dx^a}{dt} \frac{dx^b}{dt} \frac{dx^c}{dt} \frac{dx^d}{dt} = 0.$$
$$\vdots \qquad \vdots \qquad \vdots$$

The functions Γ^i_{abc}, etc., are defined by the equations,

(2·42)
$$\Gamma^i_{abc} = \frac{1}{3} P \left(\frac{\partial \Gamma^i_{ab}}{\partial x^c} - \Gamma^i_{jb} \Gamma^j_{ac} - \Gamma^i_{aj} \Gamma^j_{bc} \right),$$

(2·43)
$$\Gamma^i_{abcd} = \frac{1}{4} P \left(\frac{\partial \Gamma^i_{abc}}{\partial x^d} - \Gamma^i_{jbc} \Gamma^j_{ad} - \Gamma^i_{ajc} \Gamma^j_{bd} - \Gamma^i_{abj} \Gamma^j_{cd} \right),$$
$$\vdots \qquad\qquad \vdots \qquad\qquad \vdots$$

in which $P(\ )$ denotes the sum of the terms obtainable from the ones inside the parentheses by permuting the free subscripts cyclically. The functions Γ^i_{abc}, etc., are therefore defined so as to be symmetric in the subscripts.

If we require of a solution of (1·2) that

(2·50)
$$x^i = q^i, \quad \text{and} \quad \frac{dx^i}{dt} = a^i \quad \text{when} \quad t = 0,$$

it follows from (1·2), (2·32), etc., that the solution must be

(2·5)
$$x^i = q^i + a^i t - \frac{1}{2} (\Gamma^i_{jk})_q a^j a^k t^2 - \frac{1}{3!} (\Gamma^i_{jkl})_q a^j a^k a^l t^3 - \dots,$$

in which the subscript q indicates that the function written within the parentheses shall be evaluated at the point q.

The convergence of the series (2·5) can easily be proved by the use of dominant functions. Indeed, it is a consequence of the general existence theorem for systems of differential equations of the form

(2·6)
$$\frac{dx^i}{dt} = f^i(t, x).$$

For (1·2) is equivalent to

$$\frac{dx^i}{dt} = z^i, \quad \frac{dz^i}{dt} = -\Gamma^i_{ab} z^a z^b,$$

which is a system of first order equations of the type (2·6) in the $2n$ variables $x^1, \dots, x^n, z^1, \dots, z^n$. For the necessary existence theorem the reader is referred to Goursat's *Mathematical Analysis*, vol. II, pt. II, § 22, p. 48.

The transformation of coordinates

$$(2\cdot7) \qquad x^i - q^i = y^i - \frac{1}{2}(\Gamma^i_{ab})_q\, y^a y^b - \frac{1}{3!}(\Gamma^i_{abc})_q\, y^a y^b y^c - \dots$$

obviously transforms the curves given by (2·5) into

$$(2\cdot8) \qquad\qquad\qquad y^i = a^i t.$$

In other words, a coordinate system can be found such that the paths through the origin satisfy equations of the form (2·8). The transformation (2·7) is a special case of the transformation (13·1) of Chap. III. Hence the coordinate system y defined by (2·7) is a special type of geodesic coordinate system.

From the form of (2·8) it follows that there is one and only one curve which satisfies (1·2) and joins the origin of the coordinates y to any other point within the region of convergence of the series (2·7). Since the origin of the coordinates y can be chosen arbitrarily, it follows that the curves which satisfy (1·2) have the local property which justifies calling them a system of paths according to the definition above. It is equally obvious from (2·8) that there is one and only one path through an arbitrary point in an arbitrary direction.

3. Affine normal coordinates.

A coordinate system y will be called an *affine normal coordinate system* in case the solutions of the differential equations (1·2) which represent paths through the origin take the linear form* (2·8). The totality of affine normal coordinate systems which have a given point as origin is uniquely determined and is therefore an invariant of the point and the affine connection.

Any two affine normal coordinate systems having the same point as origin are related by a linear homogeneous transformation

$$(3\cdot1) \qquad\qquad\qquad \bar{y}^i = b^i_j y^j$$

with constant coefficients and determinant different from zero. To prove this, let us write the most general possible transformation in the form,

$$\bar{y}^i = b^i_j y^j + b^i_{jk}\, y^j y^k + \dots.$$

---

* It would not be correct to simplify this statement by saying that an affine normal coordinate system is any one in which the equations of the paths through the origin are of the form (2·8). For the paths can be represented parametrically by equations which do not satisfy differential equations of the form (1·2).

The equations (2·8), which are the solutions of the differential equations of the paths in the coordinate system y, transform into

$$(3·2) \qquad \bar{y}^i = b^i_j a^j t + b^i_{jk} a^j a^k t^2 + \ldots$$

which, according to the remark at the beginning of §2, are the solutions of the differential equations of the paths in the coordinate system $\bar{y}$. These equations will be linear of the form (2·8) only in case

$$(3·3) \qquad b^i_{jk} = 0, \quad b^i_{jkl} = 0, \ldots$$

and therefore the transformation between two affine normal coordinate systems must be of the form (3·1).

A particular affine normal coordinate system can be associated with a coordinate system x and a point q by specializing the initial conditions as follows:

$$(3·4) \qquad y = 0 \quad \text{and} \quad \frac{\partial y^i}{\partial x^j} = \delta^i_j \quad \text{when } x = q.$$

If $\bar{y}$ is the affine normal coordinate system analogously determined by an arbitrary coordinate system $\bar{x}$,

$$\frac{\partial \bar{y}^i}{\partial y^j} = \frac{\partial \bar{y}^i}{\partial \bar{x}^a} \frac{\partial \bar{x}^a}{\partial x^b} \frac{\partial x^b}{\partial y^j}.$$

Evaluating this at the origin and using the fact that any two affine normal coordinate systems having a common origin are related by a linear homogeneous transformation with constant coefficients, we find

$$(3·5) \qquad \bar{y}^i = \left(\frac{\partial \bar{x}^i}{\partial x^j} \right)_0 y^j$$

as the equations of transformation between the two affine normal coordinate systems.

4. Let us denote the components of affine connection in affine normal coordinates by Γ^{*i}_{jk}. The differential equations of the paths,

$$(4·1) \qquad \frac{d^2 y^i}{dt^2} + \Gamma^{*i}_{jk} \frac{dy^j}{dt} \frac{dy^k}{dt} = 0,$$

are satisfied by the equations

$$(4·2) \qquad y^i = a^i t,$$

in which the a's are arbitrary constants. If we substitute (4·2) in (4·1) we find

$$(4·3) \qquad \Gamma^{*i}_{jk} a^j a^k = 0$$

for all values of y on the path (4·2). Multiplying (4·3) by t^2 we find the relation

$$(4·4) \qquad \Gamma_{jk}^{*i}\, y^j y^k = 0,$$

which holds all along the path. Since the a's are arbitrary, (4·4) is satisfied for all points. If the left-hand member of (4·4) be expanded in a power series and (4·2) be substituted in, we find

$$0 = \Gamma_{jk}^{*i}\, y^j y^k = (\Gamma_{jk}^{*i})_0\, a^j a^k t^2 + \left(\frac{\partial \Gamma_{jk}^{*i}}{\partial y^l}\right)_0 a^j a^k a^l t^3 + \cdots,$$

from which we can infer at once that

$$(4·5) \qquad (\Gamma_{jk}^{*i})_0 = 0,$$

$$(4·51) \qquad \left(\frac{\partial \Gamma_{jk}^{*i}}{\partial y^l} + \frac{\partial \Gamma_{jl}^{*i}}{\partial y^k} + \frac{\partial \Gamma_{kl}^{*i}}{\partial y^j}\right)_0 = 0,$$

and in general,

$$(4·5m) \qquad S\left(\frac{\partial^m \Gamma_{jk}^{*i}}{\partial x^l \ldots \partial x^q}\right)_0 = 0,$$

in which $S(\ \ )$ stands for the sum of all the $(m+2)(m+1)/2$ terms obtainable from the one written within the parentheses by replacing the pair of subscripts jk by any pair from the set $jkl \ldots q$.

The differential equations (4·4) completely characterize the affine normal coordinates. That is to say, if we substitute for Γ_{jk}^{*i} in (4·4) the expression in terms of Γ_{jk}^{i} given by the law of transformation of the components of affine connection, we obtain a system of partial differential equations whose solutions determine the transformation from arbitrary to affine normal coordinates.

By treating the differential equations (2·32), (2·33), etc., in the same way as we have treated (1·2) above it can be proved that the identities

$$(4·6) \qquad \Gamma_{jkl}^{*i}\, y^j y^k y^l = 0,$$

$$(4·7) \qquad \Gamma_{jklm}^{*i}\, y^j y^k y^l y^m = 0,$$

$$\vdots \qquad\qquad \vdots$$

are satisfied in affine normal coordinates.

5. Affine extensions.

Covariant differentiation may be regarded (cf. § 14, Chap. III) as the process of differentiating in affine normal coordinates and evaluating at the origin. It is capable of the following immediate generalization. Given an arbitrary affine connection, the *kth affine extension* of any invariant

$$T^{ij...l}_{ab...c}$$

of the type defined in § 7, Chap. II, is the invariant

(5·1) $$T^{ij...l}_{ab...c,\, de...f}$$

whose components in any coordinate system x at any point q are given by the formula

(5·2) $$\left(T^{ij...l}_{ab...c,\, d...f}\right)_q = \left(\frac{\partial^k\, T^{*ij...l}_{ab...c}}{\partial y^d \dots \partial y^f}\right)_0$$

in which y is the affine normal coordinate system determined according to (3·4) by the affine connection and the coordinate system x at the point q, and

$$T^{*ij...l}_{ab...c}$$

are the components of T in the coordinates y. The subscripts q and 0 denote evaluation for $x = q$ and $y = 0$ respectively.

The first affine extension of a tensor is, of course, the covariant derivative. *The kth affine extension of a tensor is also a tensor.* For consider a transformation from the coordinates x to coordinates $\bar{x}$; this brings about a transformation (3·5) from the affine coordinates y with origin at q to the coordinates $\bar{y}$ determined by $\bar{x}$ at the same point. The components of T in the two normal coordinate systems are related by the equations

$$\bar{T}^{*\pi...\sigma}_{a...\gamma} = T^{*i...l}_{a...c} \left(\frac{\partial x^a}{\partial \bar{x}^a}\right)_q \cdots \left(\frac{\partial x^c}{\partial \bar{x}^\gamma}\right)_q \left(\frac{\partial \bar{x}^\pi}{\partial x^i}\right)_q \cdots \left(\frac{\partial \bar{x}^\sigma}{\partial x^l}\right)_q.$$

After differentiating k times with respect to $\bar{y}$ this yields

(5·3) $$\frac{\partial^k \bar{T}^{*\pi...\sigma}_{a...\gamma}}{\partial \bar{y}^\delta \dots \partial \bar{y}^\xi} = \frac{\partial^k\, T^{*i...l}_{a...c}}{\partial y^d \dots \partial y^f} \left(\frac{\partial x^a}{\partial \bar{x}^a}\right)_q \cdots \left(\frac{\partial x^c}{\partial \bar{x}^\gamma}\right)_q \left(\frac{\partial \bar{x}^\pi}{\partial x^i}\right)_q \cdots$$
$$\cdots \left(\frac{\partial \bar{x}^\sigma}{\partial x^l}\right)_q \left(\frac{\partial x^d}{\partial \bar{x}^\delta}\right)_q \cdots \left(\frac{\partial x^f}{\partial \bar{x}^\xi}\right)_q,$$

if we recall that

$$\frac{\partial y^i}{\partial \bar{y}^j} = \left(\frac{\partial x^i}{\partial \bar{x}^j}\right)_q.$$

The equation (5·3) states that (5·1) is a tensor with k more covariant indices than the original one.

From the definition it is clear that any affine extension is symmetric in all the indices which are introduced by the process of extension, i.e. in the indices $d, e \ldots f$ of (5·2). Moreover, the kth affine extension of the sum of any two tensors is the sum of the kth affine extensions of the two tensors. And the kth affine extension of the product of two tensors is formed by the rule for repeated differentiation of a product in elementary calculus.

6. The affine normal tensors.

The affine extensions of the affine connection itself are an important set of tensors called the *affine normal tensors*, because of their direct relation to the affine normal coordinates. That the affine extensions of the affine connection are tensors is due to the fact that between two normal coordinate systems with the same origin, the law of transformation of the Γ's is like the law of transformation of a tensor

$$\overline{\Gamma}_{jk}^{*i} = \Gamma_{bc}^{*a} \frac{\partial \overline{y}^i}{\partial y^a} \frac{\partial y^b}{\partial \overline{y}^j} \frac{\partial y^c}{\partial \overline{y}^k}.$$

Since the quantities $\partial y^i / \partial \overline{y}^j$ are constants, differentiation of this relation gives

$$\frac{\partial \overline{\Gamma}_{jk}^{*i}}{\partial \overline{y}^l} = \frac{\partial \Gamma_{bc}^{*a}}{\partial y^d} \frac{\partial \overline{y}^i}{\partial y^a} \frac{\partial y^b}{\partial \overline{y}^j} \frac{\partial y^c}{\partial \overline{y}^k} \frac{\partial y^d}{\partial \overline{y}^l},$$

from which we infer at once

$$\overline{\Gamma}_{jk,l}^i = \Gamma_{bc,d}^a \frac{\partial \overline{x}^i}{\partial x^a} \frac{\partial x^b}{\partial \overline{x}^j} \frac{\partial x^c}{\partial \overline{x}^k} \frac{\partial x^d}{\partial \overline{x}^l}.$$

In like manner we find that all the extensions,

$$\Gamma_{jk,lm}^i, \quad \Gamma_{jk,lmp}^i, \quad \Gamma_{jk,lmpq}^i, \text{ etc.}$$

are tensors.

From (4·51) and (4·5m) it follows that these tensors satisfy the identities

(6·1) $$\Gamma_{jk,l}^i + \Gamma_{kl,j}^i + \Gamma_{lj,k}^i = 0,$$

and in general

(6·2) $$S\left(\Gamma_{jk,l\ldots s}^i\right) = 0,$$

where $S(\ \ )$ has the same meaning as in (4·5m).

In order to avoid confusion with other sets of functions denoted by Γ with various arrangements of indices we shall denote the normal tensors by A with appropriate indices, so that by definition,

(6·3) $$A_{jkl}^i = \Gamma_{jk,l}^i,$$
$$A_{jklm}^i = \Gamma_{jk,lm}^i,$$
$$\vdots \qquad \vdots$$

The power series expansion for the components of affine connection in the affine normal coordinates may now be written

(6·4) $\Gamma^{*i}_{jk} = (A^{*i}_{jka})_0\, y^a + \frac{1}{2}\, (A^{*i}_{jkab})_0\, y^a y^b$

$$+ \frac{1}{3!}\, (A^{*i}_{jkabc})_0\, y^a y^b y^c + \cdots,$$

which brings out the fact that the affine connection is completely determined by the affine normal tensors.

All identities which are satisfied by all affine normal tensors are consequences of (6·2) and

(6·5) $A^{i}_{jkl\ldots s} = A^{i}_{kjl\ldots s}$ and $A^{i}_{jkab\ldots c} = A^{i}_{jkpq\ldots s}$,

in which $pq \ldots s$ stands for any permutation of $ab \ldots c$. For if at any point we assign arbitrary values to

(6·6) $A^{i}_{jkl},\ A^{i}_{jklm},\ \ldots,$

subject only to (6·2), (6·5), and the condition that (6·4) shall converge, an affine connection is determined by the series (6·4) and the co-ordinates y in which it is written are normal because they satisfy (4·4). Since the conditions of convergence are all inequalities, the quantities (6·6) can be chosen so as not to satisfy any particular assigned relation which is not a consequence of (6·5) and (6·2). This may be expressed by saying that (6·2) and (6·5) constitute a *complete set* of identities.

7. The replacement theorems.

The importance of the normal tensors is largely due to the theorem that if the components of a tensor are functions of the components of the affine connection and their first n derivatives, the components of this tensor are functions of the first n normal tensors. More precisely, if a tensor T satisfies

(7·1) $T^{i\ldots j}_{a\ldots b} = F^{i\ldots j}_{a\ldots b}\left(\Gamma^{p}_{qr},\, \dfrac{\partial \Gamma^{p}_{qr}}{\partial x^s},\, \dfrac{\partial^2 \Gamma^{p}_{qr}}{\partial x^s \partial x^t},\, \cdots\right),$

then

(7·2) $T^{i\ldots j}_{a\ldots b} = F^{i\ldots j}_{a\ldots b}\left(0,\, A^{p}_{qrs},\, A^{p}_{qrst},\, \cdots\right).$

To prove this we observe that if y is the normal coordinate system with q as origin and such that (3·4) is satisfied, then

$$F^{i\ldots j}_{a\ldots b}\left(\Gamma^{p}_{qr},\, \frac{\partial \Gamma^{p}_{qr}}{\partial x^s},\, \cdots\right) = F^{k\ldots l}_{c\ldots d}\left(\Gamma^{*p}_{qr},\, \frac{\partial \Gamma^{*p}_{qr}}{\partial y^s},\, \cdots\right)\frac{\partial y^c}{\partial x^a}\cdots\frac{\partial y^d}{\partial x^b}\frac{\partial x^i}{\partial y^k}\cdots\frac{\partial x^j}{\partial y^l}.$$

If we evaluate for $x = q$, this gives

$$(T^{i...j}_{a...b})_q = (F^{i...j}_{a...b} (0, A^p_{qrs}, A^p_{qrst}, ...))_q,$$

which, since q is arbitrary, is the same as (7·2).

8. The theorem of § 7 can obviously be extended to give the following theorem about the replacement of derivatives by affine extensions in simultaneous invariants of any basic system of invariants which includes an affine connection: If the components of a tensor T are functions of the components of an affine connection Γ and a finite number of tensors, or other invariants P, Q, ..., of the class described in § 7, Chap. II, and their first N partial derivatives, they are functions of the components of the first N normal tensors and the first N affine extensions of the invariants P, Q, etc. Moreover, if the functions mentioned in the hypothesis of the theorem are rational those mentioned in the conclusion are also rational.

So, for example, the simultaneous invariants of any quadratic differential form and any number of tensors, which are functions of the components of the basic tensors and their first N partial derivatives, are functions of the components of the basic tensors, their first N affine extensions and the first N normal tensors.

We shall prove below (§ 12) that the components of the Nth normal tensor are rational functions of the curvature tensor and its first $N - 1$ covariant derivatives and also that the components of the Nth extension of any tensor are rational functions of the components of the given tensor, the curvature tensor, and their first N repeated covariant derivatives. From this and the theorem in the section above the reduction theorems stated in § 20, Chap. III, follow at once. Thus we have finally established that all the simultaneous invariants of any basic set of invariants which includes a quadratic differential form can be generated from the basic invariants and the curvature tensor by the process of covariant differentiation.

9. The curvature tensor and the normal tensors.

By writing the formula for the curvature tensor in affine normal coordinates

$$(9·1) \qquad B^{*i}_{jkl} = \frac{\partial \Gamma^{*i}_{jk}}{\partial y^l} - \frac{\partial \Gamma^{*i}_{jl}}{\partial y^k} + \Gamma^{*b}_{jk} \Gamma^{*i}_{bl} - \Gamma^{*b}_{jl} \Gamma^{*i}_{bk}$$

and evaluating at the origin we find

$$(9·2) \qquad B^i_{jkl} = A^i_{jkl} - A^i_{jlk}.$$

Solving these equations with the aid of the identity (6·1) we find

$$(9·3) \qquad A^i_{jkl} = \tfrac{1}{3}(B^i_{jkl} + B^i_{kjl}) = \tfrac{1}{3}(2B^i_{jkl} + B^i_{ljk}).$$

By differentiating (9·2) covariantly we find that the curvature tensor and all its covariant derivatives are expressible in terms of the first normal tensor and its covariant derivatives, according to the formulas

$$B^i_{jkl,m} = A^i_{jkl,m} - A^i_{jlk,m}$$

and so on. Conversely we find by differentiating (9·3) covariantly that the first normal tensor and its first N successive covariant derivatives are expressible in terms of the curvature tensor and its first N covariant derivatives, according to the formulas,

$$A^i_{jkl,m} = \tfrac{1}{3}(B^i_{jkl,m} + B^i_{kjl,m})$$

and so on. Hence the first normal tensor and its covariant derivatives are a set of tensors fully equivalent to the curvature tensor and its covariant derivatives. We shall see below that either of these sets of tensors is equivalent to the set of all normal tensors.

10. The covariant derivative of the curvature tensor is connected with the second normal tensor by the formula

$$(10·1) \qquad B^i_{jkl,m} = A^i_{jklm} - A^i_{jlkm}.$$

This is obvious on differentiating (9·1) and evaluating at the origin of normal coordinates. Differentiating (9·1) twice in normal coordinates and evaluating at the origin gives

$$(10·2) \quad B^i_{jkl,mp} = A^i_{jklmp} - A^i_{jlkmp} + A^b_{jkm}A^i_{blp} + A^b_{jkp}A^i_{blm}$$
$$- A^b_{jlm}A^i_{bkp} - A^b_{jlp}A^i_{bkm}.$$

Differentiating (9·1) k times and evaluating at the origin will give a formula for the kth extension of the curvature tensor,

$$(10·3) \qquad B^i_{jkl,mp\ldots q} = A^i_{jklmp\ldots q} - A^i_{jlkmp\ldots q} + \ldots,$$

in which the dots represent a polynomial in the components of the first $k-1$ normal tensors.

11. The equations (10·1) together with (6·2) which in this case is

$$(11·1) \quad A^i_{jklm} + A^i_{jlkm} + A^i_{jmkl} + A^i_{kljm} + A^i_{kmjl} + A^i_{lmjk} = 0,$$

can be solved so as to yield the formula,

$$(11·2) \quad A^i_{jklm} = \tfrac{1}{6}(5B^i_{jkl,m} + 4B^i_{jlm,k} + 3B^i_{mjk,l} + 2B^i_{kml,j} + B^i_{lkm,j}).$$

In like manner the general equations (10·3) along with (6·2) yield

$$(11·3) \quad A^i_{jklmp...q} = \frac{1}{K}\left((K-1) B^i_{jkl,mp...q} + (K-2) B^i_{jlm,kp...q}\right.$$
$$\left. + (K-3) B^i_{jmp,kl...q} + ...\right) + ...$$

in which the final three dots represent a polynomial in the first $k-1$ normal tensors,

$$K = \frac{(k+3)(k+2)}{2},$$

and the subscripts of the B's are arranged according to the following scheme: $1\,2\,3$, $1\,3\,4$, ..., $1\,(m-1)\,m$, $m\,1\,2$, $2m\,(m-1)$, $2\,(m-1)$ $(m-2)$, ..., $2\,4\,3$, $3\,2\,4$, $3\,4\,5$, ..., $3\,(m-1)\,m$, $m\,3\,4$, $4m\,(m-1)$, ..., $4\,6\,5$, $5\,4\,6$, $5\,6\,7$, ..., $5\,(m-1)\,m$, ... In this scheme $m = k+3$ is the total number of subscripts for each of the B's and the letters $jkl...q$ are represented by $1\,2\,3...m$ respectively. The first and third elements of any triad are the same pair (without regard to order) as the first two elements of the next triad. The total number of triads is

$$K - 1 = \frac{m(m-1)}{2} - 1$$

because there are $m-1$ triads involving the digit 1, followed by $m-2$ other triads involving 2, and so on until the process stops with two triads involving $(m-2)$.

12. We can now prove the theorems required to complete the argument in § 8:

(A) The pth affine extension of any tensor T is a rational integral function of the first p repeated covariant derivatives of T, the curvature tensor, and the first $p-2$ repeated covariant derivatives of the curvature tensor.

(B) The pth normal tensor is a rational integral function of the curvature tensor and its first $p-1$ repeated covariant derivatives.

For the case $p=1$, (A) is a consequence of the fact that the covariant derivative is the same as the first affine extension, and (B) reduces to formula (9·3). We shall prove that they hold in general by mathematical induction. To this end we assume that both statements (A) and (B) hold for all values of p from 1 to N and prove that they hold for $p = N+1$.

Let us differentiate the formula (12·1), Chap. III, for the covariant derivative of an arbitrary tensor N times in affine normal coordinates and evaluate at the origin. The result is

$$(12\cdot1) \qquad T^{(a)}_{(\beta),j,k\dots m} = T^{(a)}_{(\beta),jk\dots m} + \dots,$$

in which the term on the left represents the Nth affine extension of the covariant derivative of T, the first tensor on the right represents the $(N+1)$st affine extension of T and the dots represent terms involving the first N affine normal tensors, the tensor T, and the first $N-1$ affine extensions of T. According to (A) for $p=1$ to N, all the affine extensions up to and including the Nth can be replaced by rational integral functions of repeated covariant derivatives of T and of the curvature tensor. According to (B) for the cases $p=1$ to N, the normal tensors in $(12\cdot1)$ can all be replaced by rational integral functions of the curvature tensor and its repeated covariant derivatives. The result of all these substitutions is that the $(N+1)$st affine extension of T, the first term on the right of $(12\cdot1)$, is expressed in the manner stated in theorem (A) for the case $p=N+1$.

In $(11\cdot3)$ we have a formula for the $(N+1)$st normal tensor in terms of the curvature tensor, its affine extensions of order 1 to N and the affine normal tensors of order 1 to N. If we substitute for the affine extensions of the curvature tensor according to theorem (A) in the cases 1 to N, and for the affine normal tensors according to theorem (B) in the cases 1 to N, we find the formula required by theorem (B) in the case $p=N+1$.

13. Affine extensions of the fundamental tensor.

Up to this point, in this chapter, we have dealt essentially with affine geometry. We come to the Riemannian geometry when we also consider the extensions of a fundamental covariant tensor g_{ij}. The first extension vanishes identically,

$$(13\cdot1) \qquad\qquad g_{ij,k} = 0,$$

as we have already seen in Chap. III (§ 15), but the higher extensions

$$g_{ij,kl}, \quad g_{ij,klm}, \quad \text{etc.}$$

are not in general zero. By the definition in § 5 they appear in the power series expansions for g_{ij} and its derivatives in normal coordinates as follows:

$$(13\cdot2) \quad g^*_{ij} = (g^*_{ij})_0 + \frac{1}{2}(g^*_{ij,ab})_0\, y^a y^b + \frac{1}{3!}(g^*_{ij,abc})_0\, y^a y^b y^c + \dots,$$

$$(13\cdot3) \quad \frac{\partial g^*_{ij}}{\partial y^k} = (g^*_{ij,ka})_0\, y^a + \frac{1}{2}(g^*_{ij,kab})_0\, y^a y^b + \dots,$$

and so on. They satisfy the identities

(13·4) $$g_{ij,ab...c} = g_{ji,ab...c},$$

and

(13·5) $$g_{ij,ab...c} = g_{ij,pq...r},$$

in which $pq ... r$ represents any permutation of $ab ... c$. These identities hold for the extensions of any symmetric tensor of the second order, but the extensions of the fundamental tensor also satisfy the identities,

(13·6) $$P(g_{ia,bc...d}) = 0,$$

in which $P(\ )$ represents the sum of all the terms obtainable from the one within the parentheses by cyclic permutations of the subscripts $abc ... d$. They also satisfy

(13·7) $$S'(g_{ab,c...di}) = 0,$$

and

(13·8) $$S(g_{ij,ab...c}) = 0,$$

in which $S'(\ )$ stands for the sum of all terms obtainable from the one within the parentheses by replacing the pair ab by every pair of indices in the set $abc ... d$ and $S(\ )$ stands for the sum of all terms obtainable from the one within the parentheses by replacing the pair ij by every pair of indices in the set $ijab ... c$. In this statement the pairs are not supposed to be ordered, so that if the number of subscripts is k, the number of terms in (13·8) is $k(k-1)/2$ and in (13·7) is $(k-1)(k-2)/2$.

14. In order to derive these identities we first prove that in the Riemannian geometry the differential equations of the paths (1·2) have a "first integral,"

(14·1) $$g_{ij} \frac{dx^i}{dt} \frac{dx^j}{dt} = \text{constant},$$

that is to say, equation (14·1) holds along any path. For if we differentiate the left-hand member of (14·1) with respect to t and replace d^2x^i/dt^2 by its value according to (1·2) we find

$$\frac{d}{dt}\left(g_{ij} \frac{dx^i}{dt} \frac{dx^j}{dt}\right) = g_{ij,k} \frac{dx^i}{dt} \frac{dx^j}{dt} \frac{dx^k}{dt},$$

which vanishes identically by (13·1).

In affine normal coordinates y, (14·1) implies the relation

$$g_{ij}^* a^i a^j = (g_{ij}^*)_0 a^i a^j$$

along any path through the origin. If we multiply by the value of t^2 determined by (4·2) this gives

$$(14·2) \qquad g_{ij}^* y^i y^j = (g_{ij}^*)_0 \, y^i y^j,$$

Since this holds for all paths through the origin it is an identity in y. It characterizes the affine normal coordinates as well as does (4·3).

If we multiply both sides of (13·2) by $y^i y^j$, sum with respect to i and j, and apply (14·2) we obtain the identity

$$(14·3) \quad 0 = \frac{1}{2}(g_{ij,ab}^*)_0 \, y^i y^j y^a y^b + \frac{1}{3!}(g_{ij,abc}^*)_0 \, y^i y^j y^a y^b y^c + \cdots$$

from which the identities (13·8) follow at once.

From (4·3) we infer

$$\left(\frac{\partial g_{ij}^*}{\partial y^k} + \frac{\partial g_{ik}^*}{\partial y^j} - \frac{\partial g_{jk}^*}{\partial y^i}\right) y^j y^k = 0,$$

or

$$(14·4) \qquad \left(\frac{2\partial g_{ij}^*}{\partial y^k} - \frac{\partial g_{jk}^*}{\partial y^i}\right) y^j y^k = 0.$$

By differentiating (14·2) we obtain

$$\frac{\partial g_{ij}^*}{\partial y^k} y^i y^j + 2 g_{ik}^* y^i - 2 (g_{ik}^*)_0 \, y^i = 0.$$

Combining this with (14·4) we have

$$(14·5) \qquad \frac{\partial g_{ik}^*}{\partial y^j} y^i y^j + g_{ik}^* y^i - (g_{ik}^*)_0 \, y^i = 0,$$

which is the same as

$$(14·6) \qquad \frac{\partial (g_{ik}^* y^i - (g_{ik}^*)_0 \, y^i)}{\partial y^j} y^j = 0.$$

Along any path we have $y^i = a^i t$, and hence (14·6) reduces to

$$(14·7) \qquad \frac{d}{dt}(g_{ik}^* y^i - (g_{ik}^*)_0 \, y^i)_0 = 0,$$

and hence

$$(14·8) \qquad g_{ik}^* y^i = (g_{ik}^*)_0 \, y^i$$

along any path through the origin. Since the path along which (14·8) has been proved to hold is arbitrary, (14·8) holds for all points. Substituting (14·8) in (14·5) we find

$$(14·9) \qquad \frac{\partial g_{ij}^*}{\partial y^k} y^i y^j = 0,$$

and when we substitute (14·9) in (14·4) we find

(14·10)
$$\frac{\partial g^{*}_{ij}}{\partial y^{k}}\, y^{j} y^{k} = 0.$$

If we multiply (13·3) by $y^{i} y^{j}$, sum with respect to i and j, and apply (14·9) we obtain

(14·11) $0 = (g^{*}_{ij,ka})_0\, y^{i} y^{j} y^{a} + \tfrac{1}{2} (g^{*}_{ij,kab})_0\, y^{i} y^{j} y^{a} y^{b} + \cdots,$

from which (13·7) follows directly. If we multiply (13·3) by $y^{j} y^{k}$, sum with respect to j and k, and apply (14·10) we get

(14·12) $0 = (g^{*}_{ij,ka})_0\, y^{j} y^{k} y^{a} + \tfrac{1}{2} (g^{*}_{ij,kab})_0\, y^{j} y^{k} y^{a} y^{b} + \cdots,$

from which (13·6) follows directly.

15. The second extension of the fundamental tensor satisfies the following identities in virtue of equations (13·4) to (13·7)

(15·1)
$$g_{ij,kl} = g_{ji,kl} = g_{ij,lk},$$
$$g_{ij,kl} + g_{ik,lj} + g_{il,jk} = 0,$$
$$g_{ij,kl} + g_{jk,il} + g_{ki,jl} = 0,$$

and these have

(15·2)
$$g_{ij,kl} = g_{kl,ij}$$

as a consequence. By writing (18·4) of Chap. III in normal coordinates and evaluating at the origin we obtain

(15·3)
$$R_{ijkl} = g_{ik,jl} - g_{il,jk},$$

and these equations can be solved so as to yield

(15·4)
$$g_{ij,kl} = \tfrac{1}{3} (R_{ikjl} + R_{jkil}).$$

From this formula or by writing (13·1) in expanded form in normal coordinates, differentiating it, and evaluating at the origin of normal coordinates we find

(15·5)
$$g_{ij,kl} = g_{ia} A^{a}_{jkl} + g_{aj} A^{a}_{ikl}.$$

By differentiating (13·1) twice in normal coordinates and evaluating at the origin of normal coordinates, we can find a formula for the second extension of the fundamental tensor in terms of the normal tensors and the fundamental tensor, and by a continuation of the process we can find the formulas for any number of extensions of the fundamental tensor. By differentiating the formula (9·2), Chap. III, for the components of affine connection in terms of the fundamental tensor and evaluating at the origin of normal coordinates, we can obtain a sequence of formulas for the normal tensors in terms of the fundamental metric tensor and its extensions. Hence this sequence of tensors can replace the sequence of normal tensors or the curvature tensor and its covariant derivatives in the reduction and equivalence theorems. The first formula of the sequence in question is

$$(15 \cdot 6) \qquad A^i_{jkl} = \tfrac{1}{2} g^{ia} (g_{aj,kl} + g_{ak,jl} - g_{jk,al}).$$

16. On account of $(15 \cdot 4)$ and $(13 \cdot 2)$ the second order terms in the expansion of $g^*_{ij} dy^i dy^j$ can now be written in the form

$$\frac{1}{3!} (R^*_{iajb} + R^*_{jaib})_0 \, dy^i dy^j y^a y^b,$$

which, because of the identity $(19 \cdot 4)$ of Chapter III,

$$R_{iajb} = \tfrac{1}{4} R_{pqrs} \delta^{pq}_{ia} \delta^{sr}_{jb}$$

reduces to $\tfrac{1}{12} (R^*_{iajb})_0 \, p^{ia} p^{jb},$
in which

$$(16 \cdot 1) \qquad p^{ia} = dy^i y^a - dy^a y^i.$$

Since any quadratic form with constant coefficients can be reduced to a sum of squares by means of a linear transformation with constant coefficients we can find a normal coordinate system in which the fundamental differential form is

$$(16 \cdot 2) \qquad g^*_{ij} dy^i dy^j = (dy^1)^2 + (dy^2)^2 + \ldots + (dy^n)^2$$
$$+ \tfrac{1}{12} (R^*_{iajb})_0 \, p^{ia} p^{jb} + \ldots.$$

This brings out clearly the manner in which the curvature tensor measures the amount by which a Riemannian geometry fails to be Euclidean. The normal coordinates of this particular type are called *Riemann's normal coordinates.* Any two such coordinate systems are of course related by an orthogonal transformation (cf. § 2, Chap. IV). The reduction to a sum of squares assumes that we are dealing with a space in which the coordinates are complex numbers; in a real space a certain number of the squares of differentials in $(16 \cdot 2)$ will in general have the coefficient $- 1$.

17. It is possible to replace the higher extensions of the fundamental tensor by a set of tensors which are symmetric in all indices after the first four and satisfy the same identities as $g_{ij,kl}$ with respect to the first four indices. Let us use the abbreviation

$$g_{ij,klab\ldots d} = \overline{ij},$$

and define the new sequence of tensors by the equations

(17·1) $(p-1)(p-2)G_{ijklab...d} = 3(ij+kl)$
$$- (ij+ik+il+jk+jl+kl),$$

in which p represents the number of subscripts. From this definition it follows directly that the new tensors are symmetric in all the subscripts after the first four and that

(17·2) $G_{ijklab...d} = G_{jiklab...d} = G_{klijab...d},$

and

(17·3) $G_{ijklab...d} + G_{ikljab...d} + G_{iljkab...d} = 0.$

From (17·1) and (15·1) and (15·2) we infer

(17·41) $G_{ijkl} = g_{ij,kl},$

(17·42) $g_{ij,kla} = G_{ijkla} + G_{ijlak} + G_{ijakl},$

and, in general, from (17·1) and (13·6), (13·7) and (13·8), we infer that

(17·4p) $g_{ij,kla...c} = S''(G_{ijkla...c})$

in which $S''(\ )$ stands for the sum of all terms obtainable from the one written within the parentheses by replacing the pair of indices kl by any pair of the set $kla...c$. If g has k subscripts there are $(k-2)(k-3)/2$ terms in $S''(\ )$. Defining

(17·5) $H_{ijkl} = (G^{*}_{ijkl})_0 + (G^{*}_{ijkla})_0\, y^a + \frac{1}{2}(G^{*}_{ijklab})\, y^a y^b + \cdots,$

we see that

$$\frac{1}{2}H_{ijkl}y^k y^l = \frac{1}{2}(g^{*}_{ij,kl})_0\, y^k y^l + \frac{1}{3!}(g^{*}_{ij,kla})_0\, y^k y^l y^a$$
$$+ \frac{1}{4!}(g^{*}_{ij,klab})_0\, y^k y^l y^a y^b + \cdots$$

on account of (17·41), (17·42), etc. Hence by reference to (13·2) we see that the fundamental quadratic form is

(17·6) $g^{*}_{ij}\, dy^i dy^j = (dy^1)^2 + \cdots + (dy^n)^2 + \frac{1}{2}H_{ijkl}\, dy^i dy^j y^k y^l.$

Moreover, by (17·2), (17·3) and (17·5)

$$H_{ijkl} = H_{jikl} = H_{ijlk} = H_{klij}$$

and

$$H_{ijkl} + H_{iklj} + H_{iljk} = 0.$$

These are the same as the symmetry relations (15·1) and (15·2) satisfied by $g_{ij,kl}$. Hence if we define

(17·7) $P_{ijkl} = H_{ikjl} - H_{iljk},$

which is analogous to (15·3), we find that

(17·8) $P_{ijkl} = -P_{jikl} = -P_{ijlk}$, and $P_{ijkl} + P_{iklj} + P_{iljk} = 0$,

and consequently,

(17·9) $P_{ijkl} = P_{klij}$.

By the argument in § 16, (17·6) transforms into

(17·10) $g_{ij}^* \, dy^i \, dy^j = (dy^1)^2 + (dy^2)^2 + \ldots + (dy^n)^2 + \tfrac{1}{12} P_{ijkl} \, p^{ij} p^{kl}$.

Corresponding to (17·7) the equations

(17·11) $K_{ijklab\ldots c} = G_{ikjlab\ldots c} - G_{iljkab\ldots c}$

define a sequence of tensors which are equivalent to the other sequences of tensors which we have been considering. These tensors are symmetric in all the indices after the first four and satisfy the same relations as the curvature tensor with respect to the first four indices.

With reference to the formulas (17·6) and (17·10) for the quadratic differential form in normal coordinates we may remark that (14·8) leads by differentiation to the following:

(17·12) $g_{ij}^* = (g_{ij}^*)_0 - \dfrac{\partial g_{ia}^*}{\partial y^j} y^a$

and

(17·13) $g_{ij}^* = (g_{ij}^*)_0 + \dfrac{1}{2} \dfrac{\partial^2 g_{ab}^*}{\partial y^i \partial y^j} y^a y^b$,

and thus in Riemann's normal coordinates, to

(17·14) $g_{ij}^* \, dy^i \, dy^j = (dy^1)^2 + \ldots + (dy^n)^2 + \dfrac{1}{2} \dfrac{\partial^2 g_{ab}^*}{\partial y^i \partial y^j} dy^i dy^j y^a y^b$.

18. Historical and general remarks.

Normal coordinates were first described by Riemann in his *Habilitationsschrift* (Part II, § 2) in geometrical terms essentially as follows: Any point can be located with respect to a fixed origin by giving the length s of the geodesic joining it to the origin and a set of quantities a^i which determine the direction of this geodesic. The latter quantities may be taken to be

$$a^i = \left(\frac{dx^i}{ds}\right)_0$$

as determined by the equations of the geodesic in an arbitrary
coordinate system. The normal coordinates then are

$$y^i = a^i s.$$

The quantities a may be transformed linearly so that the sum of
their squares shall be one and the y's then become the coordinates
referred to in § 16 as Riemann's normal coordinates.

The analytical development of this idea was given in part by
Riemann himself in his *Commentatio mathematica* (*Werke*, 2nd ed.
p. 391) and by R. Dedekind and H. Weber in *Anmerkungen* in
Riemann's collected works (2nd ed. pp. 405–409). Before this work
of Riemann's was published, however, the normal coordinates were
developed much more extensively by R. Lipschitz (*Journal für Math.*
(Crelle), vol. 70 (1869), p. 71; vol. 71 (1870), p. 274 and p. 288;
vol. 72 (1870), p. 1), who also defined normal coordinates for
differential forms of degree higher than the second. A generalization
of the normal coordinates of Riemann to a still larger class of
differential invariants was indicated by Emmy Noether in the
Göttinger Nachrichten, vol. 25 (1918). Riemann's coordinates were
used in a discussion of the equivalence theorem in the *Math. Ann.*
vol. 79 (1918), p. 289 by H. Vermeil, who also stated the formula
(17·10).

The normal coordinates have been neglected by most of the writers
on Relativity except G. D. Birkhoff (*Relativity and Modern Physics*,
Harvard Press, 1923) who used them systematically and also introduced
the tensor $g_{ij,kl}$. Normal coordinates were introduced in the affine
geometry of paths by the writer (*Proc. Nat. Ac. of Sc.* vol. 8 (1922),
p. 192) and used to define the normal tensors. The theorem of
replacement given in § 7 above, the identities (13·6) and (13·7), and
the proof that the sets of identities for the normal tensors and the
extensions of the fundamental tensors are complete were given by
T. Y. Thomas, *Math. Zeitschrift*, vol. 25 (1926), pp. 714 and 723.
We have adopted the term affine normal coordinates in the present
chapter because they are determined by the affine connection.
Other types of normal coordinates are determined by other invariants, as for example, the projective normal coordinates determined by a projective connection.

Another remark of some importance is that for many purposes it is
not necessary to know the convergence of the series (2·7) which
defines the normal coordinates. To define and discuss the kth

extension of a tensor (cf. the next section also) we use only the terms of (2·7) of orders up to and including $k + 1$. Moreover, the set of coordinate systems y defined by all transformations which agree with (2·7) up to terms of order $(k + 1)$ is a simultaneous invariant of the affine connection and the point chosen as origin. Any such coordinate system may be called a kth order affine coordinate system. For example, geodesic coordinates are first order affine coordinates.

19. Formulas for the extensions of tensors.

The fundamental formulas relating to the normal tensors and to the higher extensions of tensors in general are developed in some detail in a paper called "The Geometry of Paths," by O. Veblen and T. Y. Thomas in the *Trans. Am. Math. Soc.* vol. 25 (1923), p. 551, and a supplementary paper in the same journal, vol. 26, p. 373. In particular the reader is referred to these papers for the formulas for these tensors in arbitrary coordinates. The formula for the kth extension of any invariant is obtained by writing the law of transformation which gives the components of this invariant in normal coordinates y in terms of its components in arbitrary coordinates x, differentiating this formula k times, evaluating at the origin of normal coordinates, and eliminating the derivatives of y with respect to x by means of (2·7). For instance, to find the first extension of the affine connection, we have to treat the law of transformation of the components of affine connection (9·3), Chap. III, in the manner just described. The result is

$$A^i_{jkl} = \frac{\partial \Gamma^i_{jk}}{\partial x^l} - \Gamma^i_{jkl} - \Gamma^i_{bk} \Gamma^b_{jl} - \Gamma^i_{jb} \Gamma^b_{kl}.$$

Again, if we apply this process to the law of transformation of an arbitrary tensor of the second order, we obtain

$$T_{ij,pq} = \frac{\partial^2 T_{ij}}{\partial x^p \partial x^q} - \frac{\partial T_{ij}}{\partial x^a} \Gamma^a_{pq} - \frac{\partial T_{aj}}{\partial x^p} \Gamma^a_{iq} - \frac{\partial T_{aj}}{\partial x^q} \Gamma^a_{ip} - \frac{\partial T_{ia}}{\partial x^p} \Gamma^a_{jq}$$

$$- \frac{\partial T_{ia}}{\partial x^q} \Gamma^a_{jp} + T_{ab} (\Gamma^a_{ip} \Gamma^b_{jq} + \Gamma^a_{iq} \Gamma^b_{jp}) - T_{aj} \Gamma^a_{ipq} - T_{ia} \Gamma^a_{jpq}.$$

A more compact formula for this second extension is obtained by differentiating the formula for the covariant derivative of T_{ij} in normal coordinates and evaluating at the origin:

$$T_{ij,p,q} = T_{ij,pq} - T_{aj} A^a_{ipq} - T_{ia} A^a_{jpq}.$$

Printed in the United States
By Bookmasters